Canadian Government and Administration

A Policing Perspective

Paul F. McKenna

President, Public Safety Innovation, Inc.

With a foreword by The Honourable John C. Crosbie

PEARSON

Prentice
Hall

Toronto

Dedicated to George Grant of Terence Bay, Nova Scotia, and Glenn Gould of Uptergrove, Ontario; two peerless prodigies of the Canadian spirit.

National Library of Canada Cataloguing in Publication

McKenna, Paul F. (Paul Francis), 1952–
    Canadian government and administration: a policing perspective/Paul F. McKenna.

Includes index.
ISBN 0-13-090908-4

1. Canada—Politics and government. 2. Police power—Canada. 3. Criminal justice, Administration of—Canada. I. Title.

HV8157 M395 2004   320.471   C2003-903073-3

ISBN 0-13-090908-4

Vice President, Editorial Director: Michael J. Young
Executive Acquisitions Editor: Jessica Mosher
Assistant Editor: Andrew Simpson
Senior Marketing Manager: Judith Allen
Production Editor: Cheryl Jackson
Copy Editor: Martin Townsend
Proofreader: Claudia Forgas
Senior Production Coordinator: Patricia Ciardullo
Formatter: Heidi Palfrey
Art Director: Mary Opper
Cover and Interior Design: Anthony Leung
Cover Image: Corel Stock Photo Library/Corel Corporation

1 2 3 4 5   08 07 06 05 04

Printed and bound in Canada.

Contents

Chapter 3   **Canadian Constitutional Issues**   **63**

Chapter 4   **The Political Process**   **79**

Foreword

by The Honourable John C. Crosbie, P.C., O.C., Q.C.

I am pleased to write this foreword for Paul F. McKenna's book, his third, and final, installment in his Police Powers series. His objective, to produce an up-to-date, succinct, and solidly researched learning tool for students seeking a better understanding of police powers in Canada, is certainly accomplished, but he has also produced a first-class textbook for students or anyone interested in political science and public administration.

I endorse this book unreservedly for the use of political studies students, especially those interested in embarking on a career in policing. If everyone who enters police forces in Canada reads and absorbs the information and understanding of our democratic system contained here, we will have police forces with members alert to, and understanding of, what is proper and permissible in the process of obtaining and maintaining law and order and achieving security for those lucky enough to live in liberal democratic societies.

This is an excellent work for students planning to be police officers or any citizens interested in how our government and political system works and in understanding better the increasing complexities of life in liberal democracies.

The book discusses the fundamentals of politics and public administration, as well as the essentials of the Canadian political system and the structures and agencies of the federal system of government. It gives an overview of constitutional issues and a commentary on the essential elements of the *Canadian Charter of Rights and Freedoms*, and it delves into the mechanics of how our political system functions. It analyzes the working of political parties, party organization, and elections; describes dealing with special interest groups and lobbying, and dealing with public opinion and the role of the mass media as it relates to public safety; provides an overview of the origins and theory of public administration; looks at Canadian policy-making in policing and public safety; deals with the world of Canadian bureaucracy, federal, provincial, and municipal, and how to control the bureaucracy; outlines the issues that relate to safety, security, and surveillance; and reviews the problems of terrorism, organized crime, computer crime, and the protection of privacy versus freedom of information. Its final chapter sketches the future of Canadian politics and public administration in the context of issues of ethics, citizen engagement, accountability, restorative justice, our continuing federation, and critical issues relating to Aboriginal peoples.

Each chapter deals first with learning objectives, then after discussion of the particular subjects of that chapter, outlines questions that should be considered and discussed as a result of considering the issues raised. This is followed by a thorough listing of relevant books and references and then the listing of weblinks or the e-mail information necessary to contact a whole series of relevant associations and institutions.

As a former student in the 1950s of political science and public administration, I have no hesitation in saying this book would have been most valuable to me majoring in political science.

I recommend this fine work to intelligent Canadians interested in how their society operates politically and in the problems of modern-day living in a liberal democratic environment.

I congratulate Paul F. McKenna and heartily recommend this book to everyone interested in understanding how our system works, and should work in the future, if we are to preserve the liberal values of our democratic society and our own security.

Preface

This publication is the third, and final, installment in the Police Powers series, designed to offer an up-to-date, succinct, and solidly researched learning tool for students seeking a better understanding of police powers in Canada. Of course, while every effort has been made to ensure that the presentation of information in these chapters is current and accurate, there will inevitably be areas that quickly become dated due to the nature of Canadian law and its healthy capacity for growth and change. However, this text, in combination with the two preceding volumes, seeks to provide a resilient and reliable foundation for those working to enhance their knowledge of Canadian policing. Each chapter in this final volume offers an overview of topics that are critical to the person who wishes to embark on a career in policing or who hopes to engage in a systematic study of policing. Every effort has been made to point to additional resources and research that will broaden the student's understanding of this fascinating profession.

Chapter 1 looks at the fundamentals of political science and public administration and offers some theoretical background to this publication.

Chapter 2 considers the essentials of our Canadian political system, including the core structures and agencies of the federal system of government.

Chapter 3 offers a brief overview of certain central constitutional issues and provides some commentary on the essential elements of the *Canadian Charter of Rights and Freedoms*.

Chapter 4 delves into the political process, that is to say, the mechanics of how our political system functions, including political parties, party organization, and elections.

Chapter 5 focuses upon policing within a political context and deals with special interest groups and lobbying in the realm of public safety. Also, this chapter deals with public opinion in matters of public safety and provides some examples of political controversy involving government officials. Finally, this chapter touches upon the pivotal role of the mass media as it relates to public safety.

Chapter 6 provides an overview of the origins of public administration in Canada and outlines some of the theoretical underpinnings of this important area of study. There is some discussion of the major problems in public management, including the emergence of the "new public management" school of thought.

Chapter 7 looks at Canadian policy-making in action and concerns itself with the actual workings of the various policy communities that exist within the realm of policing and public safety.

Chapter 8 deals with the wide world of Canadian bureaucracy and addresses this topic at the federal, provincial, and municipal levels of government. We examine the available means of controlling the bureaucracy in Canada and discuss aspects of bureaucratic dysfunction.

Chapter 9 outlines the major issues in public policing that relate to safety, security, and surveillance. This includes a review of topics like terrorism, organized crime, computer crime, information and communications technology, public versus private policing, protection of privacy versus freedom of information, police educational reform, and police oversight and civilian governance.

Finally, Chapter 10 seeks to sketch a portrait of the future of Canadian politics and public administration. In this context we touch upon professional ethics, citizen engagement, accountability, restorative justice, our continuing federation, and critical issues pertaining to our Aboriginal peoples.

This textbook attempts to include a wealth of relevant background that is directly pertinent to the learning needs of the student. It aims to supply both instructor and instructed with valuable information and suitable suggestions about where to find additional research that will augment the coverage provided in these pages. As with any textbook, this one can only take the reader so far. Each chapter is equipped with clearly stated Learning Objectives that should be both realistic and relevant to the student's learning needs. The author has put a great deal of effort into assembling additional references for those readers who prefer to blaze their own research trails. Also, each chapter is equipped with a number of recommended websites that will assist in the student's further exploration of pertinent topics. As well, there are questions for discussion and several related activities that will help to deepen and strengthen the understanding gained through a reading of these chapters.

It is the purpose of this textbook to supply students with a solid grounding in the topics that are addressed within these pages. However, we are not under the illusion that this presentation will be either exhaustive or comprehensive. Students must navigate through these pages with the knowledge that there is much more to learn once the fundamentals have been mastered. The worlds of Canadian politics and public administration are not only fascinating, they are absolutely central to the essence of public policing. No one can claim to be fully conversant with the profession of policing in Canada without a fluent and inflected grasp of the topics treated in this publication.

—Paul F. McKenna

Acknowledgments

This book represents the culmination of more than five years of intensive research and writing on topics relevant to policing in Canada. When I was approached by Pearson Education Canada to undertake a series of three publications dealing with Police Powers, I did not fully appreciate how challenging this task would become. Now, at the end of this publishing cycle, I can clearly see that it was audacious in the extreme to even embark on such an enterprise. However, having said that, I am delighted to have had the privilege to undertake this assignment, whatever its flaws, for three reasons.

The first reason relates to the broad range of people I have been able to engage during the lifespan of this project and from whom I have learned so much about policing and other matters of great importance. Without their willingness to share, their thoughtfulness to respond with genuine insight, and their kindness to appreciate the author's concern for these issues, this textbook would be diminished in great measure.

The second reason relates to the enormous learning that has taken place as I reformulated, refocused, and reframed my own understanding about policing in Canada. I have never had the opportunity to examine this profession so carefully and closely. I am now more fully conscious of the nature of policing, and of its immense importance to human well being. Also, I am more deeply aware of the reality that the majority of those who become part of this world have a most profound commitment to the public good.

The third reason relates to the continuing paucity of serious, scholarly writing in the area of policing in Canada. I have heard the lament from many political scientists and professors of public administration that policing is a terribly neglected field of study. And yet, there continues to be some egregious gaps in the publishing output of our academics and professionals that leaves this territory uncharted. As a civilian in policing, and as a simple practitioner among academics, I have felt compelled to continue these efforts at assembling the data and details that serve to inform people about policing in Canada.

Because an author cannot advance without access to authoritative content, I must acknowledge those individuals who have been the source of the reasoned elaboration that provided me with the authority to say the things contained in these pages. I have no wish to assign blame if the material in this textbook is found wanting, however, if there are any merits in the presentation that follows, it is due to the generous assistance, insight, and patience of the following people. Joe Boland, President of the Royal Newfoundland Constabulary Association, for exemplifying the genuine commitment that resides within the front-line of policing. Gary Browne, Deputy Chief (retired), Royal Newfoundland Constabulary (and Member of the Police Order of Merit) who demonstrated a passion for policing, a flair for history, and a deep dedication to the profession. Dr. Barbara Bryden, who remains a leader in the learning laboratory of policing. Bill Closs, Chief of Police, Kingston, a consummate police executive who cannot say the thing which is not, and constantly reminds me that ideas have consequences. John Hagarty, Deputy Chief, Stratford Police Service, a future police leader who has shown resilience and persistence in the face of continuous change and challenge. Finally, I wish to highlight my enormous debt to Christine Silverberg, a cerebral police executive, a compassionate leader, and a dedicated decision-maker who has had a profound and lasting impact on more police personnel than anyone I have known throughout my career.

Jessica Mosher, Andrew Simpson, and Cheryl Jackson of Pearson Education Canada have offered me substantial encouragement and support throughout the process of publishing three interrelated textbooks and I am grateful for their patience and professionalism. I have been the beneficiary of excellent editing work on the part of Martin Townsend and I do appreciate the value of this subtle craft. Also, I wish to thank Claudia Forgas for her very careful proofreading of this book. Finally, as always, I reserve my deepest expressions of gratitude for my wife, Lee, and for my daughter, Kate, who, together, make the life that surrounds such efforts both magical and meaningful.

chapter one

The Fundamentals of Political Science and Public Administration

Every state is a community of some kind, and every community is established with a view to some good; for mankind always act in order to obtain that which they think good. But, if all communities aim at some good, the state or political community, which is the highest of all, and which embraces all the rest, aims at good in a greater degree than any other, and at the highest good.

—Aristotle, *Politics*, Book I, 1252a (translated by Benjamin Jowett)

Royal Newfoundland Constabulary.

Newfoundland Constabulary parade in Fort Townshend, circa 1899.

Learning Objectives

1. Identify three key perspectives on organizational theory.
2. List six basic approaches to political science and public administration.
3. Identify key elements of each of the six basic approaches referred to above.
4. Develop a familiarity with the extensive literature dealing with organizational theory.
5. Develop a specific familiarity with the Canadian literature dealing with political science and public administration.

INTRODUCTION

The Greek philosopher Aristotle observed that man is, by nature, a political animal. This was meant to imply that human beings instinctively seek to enhance their mode of existence through collective activities. It seems to mean that we are meant to live in a social environment where there is a significant level of exchange and interaction among individuals. The importance of Aristotle's apparently simple observation is that it suggests that we, as human beings, are designed to exist within structures of social organization. Of course, these social organizations may lead to both positive and negative outcomes. Certainly Aristotle did not suggest that man is, by nature, always a *successful* political animal. However, our species is naturally disposed to establish some kind of community. We are meant to share our lives with others in some manner that is directed toward our ideals of human fulfillment.

This chapter will provide a general introduction to organizational theory as it will be applied throughout this textbook. It will offer approaches to conceptual frameworks within which the reader can better understand political science and public administration as they pertain to the Canadian scene. Throughout the chapter reference will be made to some of the essential literature in this area of study, and readers will be encouraged to explore several pertinent avenues on their own initiative. Some references will connect these topics to the specific realm of policing in Canada. However, subsequent chapters will offer more considerable focus on the ways in which our various frameworks and perspectives affect the Canadian law enforcement community. Relevant examples will be drawn from the realm of Canadian policing to illustrate key points of discussion.

WHAT IS ORGANIZATIONAL THEORY?

Having suggested, following Aristotle, that human beings seek to live in organized groupings, it is helpful to subscribe to some kind of organizational theory that might guide our understanding of this reality. Theorists, over time, have looked at organizations from a number of perspectives to provide a framework for explaining how and why we gather in various structured groupings. Three of the most common perspectives may be labeled as

- the mechanistic perspective;
- the humanistic perspective and;
- the realistic perspective.

The Mechanistic Perspective

The mechanistic view of organizations sees man as a machine and treats human organizations accordingly. This entails establishing rules for the scientific management of human affairs. This perspective is focused on controlling human behaviour and maximizing efficiency. The mechanistic view came into its fullest application with the onset of the Industrial Revolution. One could say that the factory assembly line is the extreme expression of the mechanistic view of organizations, with human beings seen as adjuncts to the plant machinery. This was Frederick Winslow Taylor's perspective when he examined the world of work and production. Taylor (1947) subscribed to a belief that the application of scientific principles to human events could overcome some of society's instability and disorder.

The uncertainties, hardships, privations, and rigours of life were clear to people such as Taylor. Their drive to produce predictable results from any number of human interactions may be seen as well-intentioned, if not always successful. Frequently, the most prominent writers in this area were attempting to improve the function of business or military organizations. Clearly, police organizations may be seen as some form of hybrid of these two types of organizations, and many of the elements of police organizational life have roots in the mechanistic perspective fostered by Taylor and like-minded theorists.

One of the most important thinkers in the area of organizational theory is Max Weber (1947, 1958, 1964, and 1975). Weber's scholarly work is significant for several reasons. He is largely responsible for the formulation of ideas that cluster around the notion of "bureaucracy." As a theorist, his ideas have had a profound impact on the way we look at organizations and the people within those organizations, including civil servants. Weber's "ideal type" remains important for understanding the modern organization. The fundamental characteristics of the "ideal type" are as follows:

- All activities that are required for the achievement of a specific goal are subdivided into their most basic components. As a result, a division of labour is designed to provide expertise for specific areas of activity.

- Tasks, or activities, are undertaken in a way that allows for the application of some consistent set of rules that will contribute to reliable accomplishments. The goal of any organizational activity is to eliminate uncertainty and to introduce the highest level of predictability in the performance of tasks.

- Everyone involved within an organization is accountable to a person who is their superior in the corporate hierarchy. Each individual is responsible to that superior for their decisions. People within that organization also bear responsibility for their subordinates and their work. Authority is typically based on the capacity for reasoned elaboration, and this amounts to a kind of expertise in particular areas of activity. The final, or ultimate, authority rests with the chief official within the corporate hierarchy.

- Every official involved in an organization is expected to perform their tasks relevant to the goals of the corporate entity in an impersonal and formalistic manner. This results in a social distance being maintained between the official and any subordinates within the organization, as well as between the official and the clients of the organization. This distance is intended to foster the efficient conduct of business, unblemished by personal considerations and attachments.

- Individuals become associated with corporate entities on the basis of their personal expertise and relevant qualifications and, within the context of the bureaucratic model, will be protected against arbitrary dismissal. Advancement within the organization is based upon seniority and relevant accomplishments. Weber asserted that the notion of a "career" implies that individuals develop a fairly high degree of personal loyalty to the organization that provides them with their livelihood.

Clearly, Weber presents the bureaucratic model as something useful for the organization of human affairs and activity. He believed that this model was superior to all others in providing a stable basis for human organizations. Weber's ideas have had an immediate and profound impact on police organizations, as is evidenced by the organizational principles, based on Weber's work, that were developed by Haynes and Massie (1961, pp. 39–43) and are outlined below:

- No member of the organization should report to more than one superior.

- No superior should hold responsibility for the activities of more than five to eight subordinates.

- A superior should be prepared to delegate responsibility for routine matters to subordinates.

- Every organization should have a well-defined hierarchical structure that reveals the functional and reporting relationships that exist within the organization.

The Humanistic Perspective

The assumptions and principles of the approach to organization fostered by Taylor and Weber, which exemplify much of the mechanistic (or bureaucratic) perspective, have been questioned under what has been referred to as the humanistic school. The primary "discovery" of those who formulated the humanistic approach was that people inevitably seek out social relationships with others. This perspective came about, in part, as a result of experiments conducted in the 1920s at the Hawthorne Works, part of the Western Electric Company in Chicago. According to Gibson (1977):

> The practical application of human relations theory required careful consideration of the informal organization, work teams, and symbols that evoke worker response. Unions were viewed in a new dimension and were seen as making a contribution to effective organization rather than as the consequence of malfunctions in the organization. Participative management, employee education, junior executive boards, group decisions and industrial counseling became important means for improving the performance of workers in the organization. *(p. 8)*

The benefit of the humanistic school of thought is that it took stock of the compelling personal needs of individuals within organizations. The mechanistic approach failed to sufficiently comprehend that human beings have social needs that are powerfully active within any organizational setting and cannot be ignored, even for the sake of corporate efficiency.

In spite of the fact that employees could be compelled to work in a mechanistic manner that served the purposes of the organization, there are benefits that result from a workplace that recognizes the distinctly human needs of people. Human beings strongly desire to affiliate with others no matter what their surroundings may be and will always seek ways to achieve this goal. Abraham Maslow (1987) is another researcher who has provided us with a better understanding of the needs of individuals within organizations. Maslow's "hierarchy of needs" presents a scheme for the motivation of human beings through a progression of unsatisfied needs. That scheme is outlined in Figure 1.1.

Maslow's research led him to the belief that people are basically trustworthy and strive both to protect themselves and to govern their own behaviour. He believed that human beings are strongly motivated by affection, and that once they have taken care of the lower needs, they will increasingly seek to actualize their higher aspirations.

Physiological needs.   These are the most basic human needs. They include such things as air, water, food, sleep, and other needs of the body for sheer survival. They may be seen as essential and we, as human beings, cannot progress to the other needs until these elemental ones are, to some degree, satisfied.

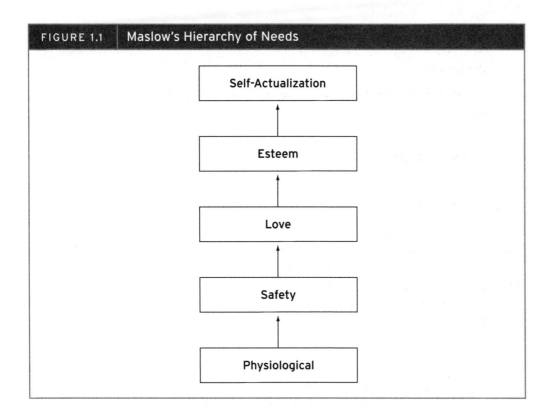

FIGURE 1.1 | Maslow's Hierarchy of Needs

Safety needs. These needs pertain to finding some measure of stability and security in our lives. Such things as family and home life are part of this level of need. These needs begin to move into the realm of psychological requirements.

Love needs. As human beings, we have a strong desire for affection and strive to attain some sense of belonging. As a result, we find ourselves in various kinds of groupings that cater to these needs, for example, religious affiliations, work groups, and families. As social beings, we strive for contact with others to gratify our desire for affection and love.

Esteem needs. Maslow identified two levels of esteem: self-esteem, which derives from a sense of our own accomplishments or competence in some given area of activity; and the esteem that comes from the recognition and attention of others.

Self-actualization needs. At the top of Maslow's ladder is the need for individuals to achieve some sense of personal fulfillment. This may never be accomplished; however, when the previous stages have been adequately met, we move into a realm where we seek to realize our fullest potential as human beings and accomplish something special in our lives.

The issue of control and the need for supervision to ensure productivity and advancement of organizational goals continues to be important for the humanistic school of organization. However, these theorists warn against treating human beings as cogs within the wheels of organizational machinery. Several classic works on organization by Merton (1940), Gouldner (1954), and March and Simon (1958) are important for readers who wish to appreciate the humanistic perspective more fully.

The Realistic Perspective

A criticism of the humanistic school came into being that not only questioned its assumptions, but also noted that this perspective has often been used to manipulate workers to accomplish organizational goals. For example, several authors, including the prolific Peter Drucker (1962, 1969, 1993a), have pointed out that corporations (both public and private) have no right to claim total loyalty from their employees.

The realistic perspective offers something of a synthesis of the two earlier models discussed above. It is one that views modern organizations as complex systems. It recognizes the need for appropriate controls (e.g., administrative, financial, procedural, and operational) as well as supervision that will ensure the productivity and advancement of organizational goals and objectives. However, there is an additional understanding that individuals constitute one of the key variables within the organizational system and that conflict will be an inevitable element of organizational life.

The realistic perspective recognizes that the concept of self-actualization, as formulated by Maslow and outlined above, is important to human beings yet maintains that it is highly unlikely that true self-actualization will occur within an organizational context. The work of Chris Argyris (1957, 1962, 1964, 1973, 1993) examines the role of the individual within an organizational setting and provides valuable insight into how modern, innovative organizations ought to function. The fact that people in public and private corporations may be compelled to contribute to the achievement of organizational goals through their individual efforts is clearly recognized. The specific need of individuals for group cohesion and identity is also seen as important to effective organizational life. Argyris observes that the inevitable tension between the overall goals of any given organization and the more profound, and individualistic, self-actualization goals of discrete human beings tends to complicate the performance of modern organizations but is what makes the study of such matters so fascinating and important.

INTRODUCTION TO POLITICAL SCIENCE AND PUBLIC ADMINISTRATION

Having established several basic theoretical perspectives on organizations, we turn our attention to the realms of political science and public administration. These are areas of critical importance to our understanding of policing in Canada because they have such a penetrating impact on what the police do, and the manner in which policing is carried out in our society.

While political science and public administration may best be viewed as distinct disciplines, they share a number of common features that are meaningful for those who function within police organizations. Furthermore, there are various approaches to these two disciplines that can be of value when attempting to arrive at a coherent understanding that is applicable to both areas of activity and study. It is important to have a degree of flexibility in examining these two related areas in order to arrive at a level of comprehension that is neither rigid nor wholly unstructured.

What Is Political Science?

Political science may best be seen as the systematic study of politics and those matters that pertain to the acquisition and disposition of power. It is based on the Greek word for the city, *polis* (or "πολις"), and concerns itself with those issues that involve the city. From Dunner (1970):

[P]olitical science has often been defined as the systematic study of the state and of the processes governing its internal and external relations. *(p. xv)*

In subdividing this large topical area, the following categories may be of interest:

- *political theory,* or political philosophy, which includes the formulation and discussion of ancient and modern political ideas;

- *government,* including the consideration of constitutions, national governments, regional and local governments, public administration, the economic and social functions of government, and comparative political institutions;

- *parties, groups, and public opinion,* including the full range of political parties, special interest and pressure groups and associations, aspects of citizen participation in government and administration, and public opinion; and

- *international relations,* including the study of international politics, international organizations and administration, and international law.

In an effort to simplify some of this broad study, it may be helpful to consider the elements of the ancient world's understanding of human knowledge, as outlined in Figure 1.2.

The ancients viewed logic as a necessary prelude to the contemplation of science, or philosophy. The study of logic has its importance today; however, it does not form a preliminary discipline in our conventional curriculum. Yet the ancients saw logic as an essential tool for any skills in reasoning, and the Platonic dialogues may be seen as a remarkable and extended tribute to the master logician, Socrates, the man whom the gods called the wisest of men. The ancients understood science to be essentially synonymous with philosophy, and these terms simply applied to the concept of "knowledge" as we conventionally understand it today. The considerable distinction between science and philosophy is a product of the modern world (Grant, 1986). The areas that branched off from science or philosophy could be either theoretical or practical. Within the realm of the theoretical were studies in physics, metaphysics, and mathematics. Within the realm of the practical were studies in politics, ethics, and economics.

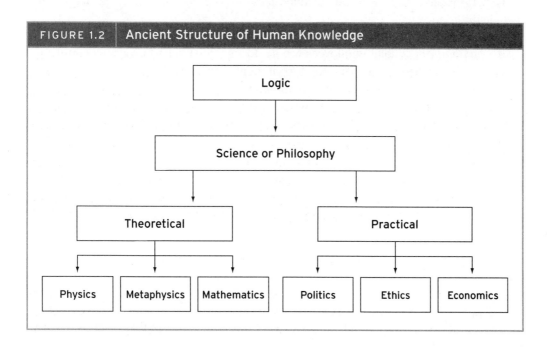

FIGURE 1.2   Ancient Structure of Human Knowledge

Politics would deal with all matters pertaining to actions and deliberations within the state (or *polis*) and would include the making of laws; the responsibilities, obligations, and rights of citizens; the details of war and peace; and relations with other cities. Ethics would concern itself with right action (justice) and the relationship between citizens as human beings (Anastaplo, 1975). Economics, as the ancients understood this term, pertained to skill in the management of the household and was seen to be an important art for citizens that required considerable knowledge and application.

What Is Public Administration?

Public administration may be seen as the theory and practice associated with the management of public affairs. It includes the consideration of matters that pertain to public affairs at the international, national, provincial, regional, and local municipal levels. It is often seen as a subset of political science. However, as Kernaghan (1977) indicates:

> The scope of public administration is now so broad and its subject matter so heterogeneous that it cannot realistically be considered solely as a subdiscipline of political science. It is clear that appropriate settings for education in public administration exist outside the structural framework of political science departments and that a variety of organizational patterns have evolved. *(p. 40)*

Others view public administration as a specific form of administrative study. Accordingly, Kernaghan (1977) makes the following observation:

> Those who view public administration as a generic process contend that the study of administration may be most fruitfully pursued by dividing the general fields into public, business, hospital, educational, police, or other types of administration. To proponents of this view, the similarities between these various forms of administration are greater than their differences. This belief has led to the establishment of "professional" schools or faculties of administrative studies. *(p. 37)*

Generally, the discipline of public administration is guided by certain principles that include the following:

Accountability.   Public employees are held to some standard of accountability for the quality of their advice, and for carrying out their assigned duties and seeking to achieve policy and program objectives within the framework of law, recognizing existing constraints, taking direction from their superiors, and working within the limits of the authority and resources at their disposal. Public employees are also accountable on a day-to-day basis to their superiors within the bureaucracy for their own actions and the actions of their subordinates. They owe their primary duty, however, to their political superiors. They are indirectly accountable to the appropriate legislature or council and to the public through their political superiors. Public servants have a further responsibility to report any violation of the law to the appropriate authorities.

Service to the public.   Public employees will provide quality service to the public in a manner that is courteous, equitable, efficient, and effective. Public employees are expected to be sensitive and responsive to the changing needs, wishes, and rights of the public while respecting the legal and constitutional framework within which service to the public is provided. In order to promote excellence in public service, public employees have a responsibility to maintain and improve their own functional competence, as well as that of their colleagues.

The public interest.   Public employees are expected to resolve any conflicts that might arise between their personal or private interests and their official duties in favour of the public interest. Public employees seek to serve the public interest by upholding both the letter and the spirit of the laws established by the appropriate legislature or council and of the regulations and directions made under these laws.

Political neutrality.   Public employees must be sensitive to the political process and knowledgeable about the laws and traditions regarding political neutrality that are applicable to their employment circumstances. Public employees are responsible for providing forthright and objective advice to, and for carrying out the directions of, their political superiors. Public employees must execute government decisions loyally without regard to the party or persons in power and without reference to their personal opinions.

Political rights.   Public employees must be able to exercise the fullest possible measure of political rights compatible with laws, regulations, and conventions in place to preserve the political neutrality of the public service. They have a responsibility to avoid participation in partisan politics that is likely to impair the political neutrality of the public service or the perception of that neutrality. In exchange, public servants will not be compelled to engage in partisan political activities or be subjected to threats or discrimination for refusing to engage in such activities. Public employees will not express personal views on matters of political controversy or on government policy or administration when such comment is likely to jeopardize public confidence in the objective and efficient performance of their duties. It is the responsibility of public servants to seek approval from the appropriate governmental authority whenever they are uncertain as to the legality or propriety of expressing their personal views.

Conflict of interest.   Public employees must not engage in any business or other transaction or have any financial or other personal interest that is, or may appear to be, incompatible with the performance of their official public duties. They should not, in the performance of their official duties, seek personal or private gain by granting preferential treatment to any persons. Public employees should neither solicit nor, unless duly authorized, accept transfers of economic value from persons with whom they have contact in their official capacity. Public employees will not use, or permit the use of, government property of any kind for activities not associated with the performance of their official duties, unless they are authorized to do so. They should not seek or obtain personal gain from the use of information acquired during the course of their official duties which is not generally available to the public.

Confidentiality.   Public employees will not disclose to any member of the public any secret or confidential information acquired as a result of their official position. Adhering to the boundaries established by law and propriety, public employees will be sensitive and responsive to the needs of the public, the news media, and legislators for information on and explanation of the content and administration of government policies and programs.

Discrimination and harassment.   All public employees have a duty to treat members of the public and one another fairly and to ensure that their work environment is free from discrimination and harassment. (Adapted from Statement of Principles, Institute of Public Administration of Canada.)

APPROACHES TO THE STUDY OF POLITICAL SCIENCE AND PUBLIC ADMINISTRATION

As with our earlier discussion of organizational theory, it is useful to provide some specific focus for our examination of political science and public administration. Rather than establish a rigid theoretical framework for this purpose, this textbook will attempt to incorporate several ways of looking at these areas of study. This will allow the reader a considerable degree of flexibility when examining the impact of political science and public administration on policing in Canada. Such an approach underscores the belief that no single perspective offers a comprehensive insight into the interrelationships that exist among these topics.

A number of different approaches may be taken to the study of either political science or public administration. The following six basic approaches have some claim on our attention:

- the systems approach

- the pluralist approach

- the public choice approach

- the elite approach

- the neo-Marxist approach

- the state-centred approach

The Systems Approach

The systems approach views political entities as complex structures that contain elements that are mutually dependent and which interact on various levels. Of course, the individuals that make up these entities become one of the key variables that influence the nature of these interactions. Linkages within the political entity are established through processes of communication, decision-making, structured roles, and other means of control. The systems approach allows for a structural understanding of the political realm that features the pivotal role of human actors within the system.

Rand Dyck (1993) uses a systems analysis approach in his examination of Canadian politics. It is a valuable framework for the study of politics and government that is derived from the work of David Easton, a Canadian political scientist.

> Systems theory is closely related to structural-functionalism, of which Gabriel Almond was a pioneer. This line of analysis begins with the identification of the functions that are or must be performed in any political system, and then searches for the structures that perform them. Almond postulated that seven basic functions are performed in any political system: interest articulation, interest aggregation, political socialization and recruitment, political communication, rule making, rule application, and rule adjudication. *(p. 6)*

The Pluralist Approach

This is an ideal of democratic practice that perceives that individuals are able to take advantage of several different resources in the process of participating in the democracy.

This is a theory that was most fully developed by the American political scientist Robert Dahl. It is a theory that sees groups coming together under various conditions to make demands on government authorities with respect to a diverse range of policy issues. There is an assumption within this approach that group action is more effective than individual action, and this is particularly the case with specific pressure groups that have been created to pursue a specific policy objective. For example, victims' groups, gun legislation lobbyists, environmental groups, and poverty activists are all collectives of individuals who share a common purpose that operates to influence public policy on some single issue. The work of A. Paul Pross (1992) provides an excellent example of the pluralist approach and the impact of groups on Canadian politics.

Dyck (1993) summarizes the pluralist approach as follows:

- Power is widely dispersed in the political system and not monopolized by any state or corporate elite.

- Individuals are free to employ a variety of resources at their disposal and to organize whatever groups they wish in order to back their demands upon the authorities.

- The decisions made by the authorities are basically compromises among the various competing interests that articulate their demands.

- Different policy areas are characterized by different individuals and groups making demands upon different authorities.

- Pressure group activity is increasingly replacing individual and party activity in the political system. *(p. 8)*

The Public Choice Approach

This model is based on the assumption that citizens, who are seen to be rational, self-interested players in the democratic process, will support political parties whose policies provide the greatest benefit to them as voters. This approach views politics as an exercise in opportunity, with politicians seeking votes at virtually any cost, in order to secure election. This, of course, is an understanding that can lead to a fair degree of cynicism about the political process. However, there is a considerable amount of reality in this viewpoint, and it may be seen as being in many ways utilitarian (that is, achieving the greatest good for the greatest number). Dyck (1993) usefully summarizes this approach:

- Politicians and voters act in a rational, self-interested fashion in which politicians make promises in return for votes.

- Politicians and parties generally adopt policies that will get themselves elected, and other things being equal, they respond to those interests representing the largest number of votes.

- Since it is a waste of effort to appeal to committed supporters or opponents, however, politicians concentrate on marginal, undecided, or strategically located voters.

- Politicians try to maximize publicity of their successes and minimize their failures, take credit for good things and blame others for the bad, and manipulate the timing of positive and negative decisions.

- A similar bargaining process goes on at other points in the political system, such as between politicians and the bureaucracy, the authorities and pressure groups, and the authorities and the media. *(p. 9)*

The Elite Approach

This approach takes the view that small groups of people hold a disproportionate amount of power within the political system. These elites, necessarily, include those people who occupy elected office, as well as those who possess socio-economic advantages beyond those of ordinary citizens. For example, it is asserted that those within the private sector, and who hold significant economic power, are also likely to wield considerable political influence as a direct result of that power. Many writers have focused on the power of elite groups within the Canadian political fabric (Porter, 1965; Presthus, 1965; and Olsen, 1980). The concept of a concentrated Canadian "establishment" is consistent with the elite approach.

The notion of "elite accommodation" deals with the interaction of various elite groups that may function within government configurations or or may involve relationships between elite groups within the private or public sectors. This view is carefully articulated by Presthus (1973). Beyond the influence of elite groups, there are many cleavages in Canadian political life, including ethnic, religious, regional, and economic, among others, ensuring that accommodation becomes an important means of arriving at workable compromises for public policy and programs. Each of these cleavages may be seen to contain its own elite group that wields a disproportionate amount of power and influence within the Canadian political system.

The Neo-Marxist Approach

Clearly, this approach is derivative of the political and social thinking of Karl Marx. In spite of the apparent repudiation of Marxist economic theory symbolized by the collapse of the former Soviet Union and the ongoing ideological transformation of Eastern Europe, there remains much in Marx's writings that is important to any understanding of the capitalist system or class structure. Those who subscribe to a neo-Marxist perspective believe that the political elite within any given jurisdiction is guided by the capitalist elite, which is driven by the need to accumulate capital. There are many Canadian political scientists who view our political and economic landscape from a neo-Marxist perspective (Clement, 1975; Panitch, 1977), and this approach remains a vibrant mode of analysis for Canadian politics.

The State-Centred Approach

This approach argues that those who hold political authority within the state are essentially autonomous of the other influences mentioned above. These authorities, both elected and bureaucratic, formulate public policy without much substantial reference to public demands, and they often operate in opposition to those demands. This theoretical perspective has been most clearly developed by Eric Nordlinger (1981). The importance of experts within various policy fields is worth considering when looking at the development of public guidelines and standards of practice consistent with the state-centred approach. Here the bureaucratic layer that functions below the level of elected officials is pivotal. Dyck's (1993) perspective is again instructive in this instance:

> While members [of] the political executive (the prime minister and cabinet in Canada) have the right to make major governmental decisions (subject to some kind of accountability to Parliament), some theorists prefer to emphasize the influence of the bureaucracy within the apparatus of the state. They see the bureaucracy as the most important part of the policy-making process because neither legislators nor cabinet ministers can possibly understand the detail of complex modern issues, and are therefore content to be advised by their experts. Politicians may still have a role to play in identifying problems and determining priorities, but once that is done they are in the hands of their advisers in finding ways to proceed. *(p. 13)*

Each of the six approaches outlined above has something valuable to offer in the careful analysis or assessment of Canadian politics. In the following chapters we will consider various aspects of the Canadian political system, as well as elements of Canadian public administration. In relating these topics to policing, it is worthwhile to have a number of vantage points from which to view these subjects. The political realm is particularly complex and, therefore, it is valuable to be in a position to assess the political dimension from several perspectives. This will permit readers to broaden and deepen their appreciation of the impact of things political on the world of Canadian policing.

CONCLUSION

Because all police departments represent some form of organization, it is essential that we have an effective understanding of organizational theory. This chapter has offered three basic perspectives that may inform our appreciation of organizations. However, the literature dealing with this area of concern is both complex and ever-changing. It is fundamentally important to appreciate the organizational setting of policing in Canada, and this justifies the careful study of the relevant theoretical and practical analysis in this area.

Also, as significant institutions that inhabit a landscape that is deeply and extensively influenced by the political realm, Canadian police organizations must be understood in the context of the science (or philosophy) of politics. Every police service in Canada is guided by at least one level (and often several levels) of government. Therefore, the serious student, or conscientious practitioner, of policing in Canada must be highly familiar with certain elements of political science.

Finally, as a civil service *par excellence,* policing is continuously responsible to the public good. Because police services draw, almost exclusively, upon the public purse for their funding and support, it is essential that there be a close reading of the principles of public administration and a strict adherence to the highest standards of stewardship.

REFERENCES

Adie, Robert and Paul Thomas (1987). *Canadian public administration. 2nd ed.* Scarborough, Ont. : Prentice-Hall Canada.

Anastaplo, George (1975). *Human being and citizen: essays on virtue, freedom, and the common good.* Chicago : Swallow Press Inc.

Argyris, Chris (1957). *Personality and organization: the conflict between system and the individual.* New York : Harper & Row.

———— (1962). *Interpersonal competence and organizational effectiveness.* Homewood, Ill. : Dorsey Press.

———— et al. (1962). *Social science approaches to business behavior.* Homewood, Ill. : Dorsey Press.

———— (1964). *Integrating the individual and the organization.* New York : Wiley.

———— (1973). *On organizations of the future.* Beverly Hills, Calif. : Sage Publications.

———— (1993). *Knowledge for action: a guide to overcoming barriers to organizational change.* San Francisco : Jossey-Bass Publishers.

———— and Donald A. Schön (1977). *Theory in practice: increasing professional effectiveness.* San Francisco : Jossey-Bass Publishers.

Aucoin, Peter (1996). "Political science and democratic governance." *Canadian Journal of Political Science,* Vol. 29, No. 4, pp. 643–660.

Bealey, Frank (1999). *The Blackwell dictionary of political science: a user's guide to its terms.* Malden, Mass. : Blackwell Publishers.

Booth, Geoffrey and Dennis Roughley (2001). *Canadian political structure and public administration: a law enforcement perspective.* Toronto : Emond Montgomery Publications.

Brooks, Stephen and Andrew Stritch (1991). *Business and government in Canada.* Scarborough : Prentice-Hall Canada.

Clement, Wallace (1975). *The Canadian corporate elite.* Toronto : McClelland and Stewart.

Cooper, Cary L. and Chris Argyris (eds.) (1997). *The Blackwell encyclopedic dictionary of management.* Cambridge, Mass. : Blackwell.

Dahl, Robert (1967). *Pluralist democracy in the United States.* Chicago : Rand McNally.

Drucker, Peter F. (1962). *The new society: the anatomy of industrial order.* New York : Harper & Row.

———— (1969). *The age of discontinuity: guidelines to our changing society.* New York : Harper & Row.

———— (1993a). *Concept of the corporation.* New Brunswick, N.J. : Transaction Publishers.

———— (1993b). *Post capitalist society.* Oxford : Butterworth-Heinemann.

Dunner, Joseph (1970). *Dictionary of political science.* Totowa, N.J. : Littlefield, Adams.

Dyck, Rand (1993). *Canadian politics: critical approaches.* Toronto : Nelson Canada.

Easton, David (1965). *A systems analysis of political life.* New York : John Wiley & Sons.

Fleming, James (1991). *Circles of power: the most influential people in Canada.* Toronto : Doubleday.

Gibson, James L. (1977). "Organization theory and the nature of man." In Kernaghan, Kenneth (ed.). *Public administration in Canada: selected readings. 3rd ed.* Toronto : Methuen.

Gouldner, Alvin W. (1954). *Patterns of industrial bureaucracy.* New York : The Free Press of Glencoe.

Grant, George P. (1965). *Lament for a nation: the defeat of Canadian nationalism.* Toronto : McClelland and Stewart Limited.

———— (1966). *Philosophy in the mass age.* Toronto : Copp Clark Publishing.

———— (1969). *Technology and empire: perspectives on North America*. Toronto : House of Anansi.

———— (1986). "Thinking about technology." In *Technology & Justice*. Toronto : Anansi.

Guy, James John (1997). *Expanding our political horizons: readings in Canadian politics and government*. Toronto : Harcourt Brace Canada.

Haynes, W. Warren and Joseph L. Massie (1961). *Management*. Englewood Cliffs, N.J. : Prentice-Hall, Inc.

Heimanson, Rudolph (1967). *Dictionary of political science and law*. Dobbs Ferry, N.Y. : Oceana Publications.

Hodgetts, J.E. and D.C. Corbett (eds.) (1960). *Canadian public administration*. Toronto : Macmillan.

Janis, Irving (1989). *Crucial decisions: leadership in policymaking and crisis management*. New York : The Free Press.

Kernaghan, Kenneth (ed.) (1977). *Public administration in Canada: selected readings. 3rd ed*. Toronto : Methuen.

Kernaghan, Kenneth and David Siegel (1991). *Public administration in Canada. 2nd ed*. Scarborough, Ont. : Nelson Canada.

March, James G. and Herbert A. Simon (1958). *Organizations*. New York : John Wiley and Sons.

Maslow, Abraham H. (1987). *Motivation and personality. With new material by Ruth Cox and Robert Frager. 3rd ed*. New York : Harper and Row.

Merton, Robert K. (1940). "Bureaucratic structure and personality." *Social Forces,* Vol. 18, pp. 560–568.

Nordlinger, Eric A. (1981). *On the autonomy of the democratic state*. Cambridge, Mass. : Harvard University Press.

Olsen, Dennis (1980). *The state elite*. Toronto : McClelland & Stewart.

Panitch, Leo (1977). *The Canadian state*. Toronto : University of Toronto Press.

Porter, John A. (1957). *The economic elite and the social structure in Canada*. Toronto : S.n.

———— (1965). *The vertical mosaic*. Toronto : University of Toronto Press.

Presthus, Robert V. (1965). *Behavioral approaches to public administration*. Tuscaloosa : University of Alabama Press.

———— (1973). *Elite accommodation in Canadian politics*. Toronto : Macmillan.

Pross, A. Paul (1992). *Group politics and public policy. 2nd ed*. Toronto : Oxford University Press.

Taylor, Frederick Winslow (1947). *Scientific management, comprising shop management, the principles of scientific management [and] testimony before the Special House Committee*. New York : Harper.

Tindal, C. Richard (1997). *A citizen's guide to government*. Toronto : McGraw-Hill Ryerson Limited.

Weber, Max (1947). *The theory of social and economic organizations*. Translated by A.M. Henderson and Talcott Parsons. New York : Oxford University Press.

———— (1958). *The Protestant ethic and the spirit of capitalism*. Translated by Talcott Parsons with a foreword by R.H. Tawney. New York : Scribner.

———— (1964). *The methodology of the social sciences*. Translated and edited by Edward A. Shils and Henry A. Finch. New York : Free Press of Glencoe.

———— (1975). *Max Weber: the interpretation of social reality*. Edited with an introduction by J.E.T. Eldridge. New York : Scribner.

Wilson, James Q. (1968). *Varieties of police behavior: the management of law & order in eight communities*. Cambridge, Mass. : Harvard University Press.

WEBLINKS

Canadian Political Science Association

This organization was founded in 1913 and functions to encourage and develop political science and its relationship with other disciplines. It publishes the *Canadian Journal of Political Science,* as well as books and monographs on topics relating to political science.

www.cpsa-acsp.ca

Institute of Public Administration of Canada

This institute, founded in 1927, is Canada's leading organization for public management. It deals with the theory and practice of public administration and publishes extensively in this field. Of particular note is the scholarly journal *Canadian Public Administration,* which examines the structures, processes, outputs, and outcomes of public policy at all levels of Canadian government. IPAC also publishes *Public Sector Management Magazine* and an ongoing series of monographs on Canadian public administration that are important for scholars, practitioners, and students in this area of study.

www.ipaciapc.ca

QUESTIONS FOR CONSIDERATION
AND DISCUSSION

1. Which of the three perspectives on organizational theory applies most consistently to police organizations in Canada?

    _____

    _____

    _____

2. What are some differences between political science and public administration?

    _____

    _____

    _____

    _____

    _____

    _____

    _____

    _____

    _____

    _____

    _____

    _____

3. To what degree does the realm of political science affect the development of policing in Canada?

    _____

    _____

    _____

    _____

4. To what degree does public administration affect the development of policing in Canada?

_____

_____

_____

_____

_____

5. Which of the approaches to political science and/or public administration outlined in this chapter is best suited to progressive policing in Canada? Which of these approaches is least suited to progressive policing?

_____

_____

_____

_____

6. What elite groups can you identify in Canada that might have significant political influence?

_____

_____

_____

_____

_____

7. What pressure groups are you aware of in Canada that might have an interest in policing or public safety issues?

_____

_____

_____

_____

_____

RELATED ACTIVITIES

1. Consider the impact of political science on policing in Canada. Compare your findings with the impact of political science on policing in the United States or some other jurisdiction (e.g., Great Britain, France, Russia).

2. Consider the impact of public administration on policing in Canada. Compare your findings with the impact of public administration in the United States or some other jurisdiction (e.g., Great Britain, France, Russia).

3. Provide a brief analysis of Canadian politics that is expressed in the context of the pluralist approach. Provide a similar analysis of Canadian politics that adopts the other approaches: public choice, elite, neo-Marxist, and state-centred.

The Essentials of the Canadian Political System

A highly civilised society would be one in which one could be perfectly and conscientiously virtuous, think seriously about everything one did, and not be insulted.

—Simone Weil, *Lectures on philosophy* (translated by Hugh Price)

Royal Newfoundland Constabulary.

Newfoundland Constabulary Chief of Police and wife meet with Her Majesty Queen Elizabeth and the Prince of Wales in St. John's, 1956.

Learning Objectives

1. Identify the basic elements of Canadian political culture.
2. Identify the core institutions of the Canadian political system.
3. List six major distinctions that characterize Canadian politics.
4. Describe the major aspects of the division of powers between the federal and provincial governments in Canada.
5. Describe the role of municipalities in the Canadian political system.
6. List the key features of the Canadian political structure.
7. Identify the ministries and departments that have responsibilities for public safety and security in the ten provinces and three territories in Canada.
8. Identify the five major political ideologies that exist within Canadian politics.

INTRODUCTION

In this chapter we will consider the essential elements of the Canadian political system. We'll begin with some review of the nature of the Canadian political culture and the core institutions that provide the infrastructure for our system of government. This section will provide a general assessment of those things that afford us a unique identity as a country and which provide us with our common bond as citizens. Next, there will be some consideration given to the various forms of distinction within Canada, or what Dyck (1993) refers to as "cleavages." Here there will be some opportunity to inquire into what distinguishes Canadians from one another across the nation.

A very brief consideration of the concept of federalism will be offered as it represents the distinctive form of Canadian government and is pivotal to a clear understanding of our political dynamic. There will be some examination of the issues that affect the three major levels of government in Canada: federal, provincial, and municipal. Here there will only be a cursory review in light of the complexity of this topic, as well as the fact that the important constitutional aspects of this topic will be discussed in more detail in Chapter 3 ("Canadian Constitutional Issues").

Next there will be an outline of the major components of the Canadian political system at the federal level, namely the Crown, the Senate, the House of Commons, the Office of the Prime Minister, the Cabinet, the judiciary, and the federal bureaucracy. Each of these elements will be reviewed in outline, and reference will be made to the existing literature that affords a more detailed analysis of these important areas of study. Chapter 8 ("Canadian Government Bureaucracy") will take up the important theme of bureaucracy in more detail.

We will briefly consider each of the provincial and territorial jurisdictions to better understand their individual responsibilities for the provision of public safety, policing, and law enforcement services within their boundaries. Additional information will be provided in other chapters dealing with specific aspects of the provincial and/or municipal engagement in public safety and law enforcement. However, this initial treatment will provide a suitable framework for subsequent discussion.

Finally, we will look at the five major political ideologies that currently guide Canadian politics. Each of these ideologies has had some influence on the development of political life in Canada and continues to inform many of our ideas about government. Again, readers will be introduced to the extensive literature that exists so they can explore this area more extensively. A more specific discussion of the major political parties in existence in Canada will be reserved for Chapter 4.

CANADIAN POLITICAL CULTURE

According to Bell and Tepperman (1979), political culture may be defined as follows:

> Political culture consists of the ideas, assumptions, values, and beliefs that condition political action. It affects the ways we use politics: the kinds of social problems we address and the solutions we attempt. For political culture serves as a filter or lens through which political actors view the world; it influences what they perceive as social problems and how they react to them. *(p. 246)*

The most general notion that one should clearly understand about our political culture is that Canada is a democracy. Very simply stated, democracy means rule by the people (the Greek word for "the people" is *demos*). However, there are many forms that democracy may take in reality. One need only compare the People's Republic of China under Chairman Mao with Athenian democracy during the time of Demosthenes to realize that this concept is exceedingly fluid and flexible. This is clearly a complex area that requires considerable attention. Accordingly, for the purpose of this study, we will concentrate on the immediate aspects pertaining to the most relevant points that characterize Canadian democracy.

We are often able to explain what we are, politically, by pointing to what we believe we're not. Through this method we operate by comparison. For example, we tend to feel quite strongly that our Canadian democratic identity is different from that of the United States. We pride ourselves on being a nation that is geared toward "peace, order, and good government." We are quick to point to our sustained international reputation as peacekeepers in the world arena. We often characterize our country by pointing to the painstaking and patient work of the Fathers of Confederation. Their efforts tended to feature compromise and consensus. The American Founding Fathers presided over a revolutionary founding that saw bloodshed as well as considerable political reflection. One central fact of our democratic tradition, and one that we share with our neighbours to the south, is the existence of free elections supported by the right of citizens to vote. Hill (1974) outlines the major features of a representative government that gives expression to democratic principles as follows:

> Democratic theory, as it has evolved from classical liberalism in Victorian times, centres on representative government to fulfil its aims of liberty, equality and fraternity. The basic tenets of representative institutions are: free elections; majority rule; protection of minorities, subject to the majority's final say; and the assumption that government operates on the basis of widespread discussion and a responsiveness to an informed public opinion. *(pp. 20–21)*

Popular Sovereignty   Popular sovereignty is a term that is usefully applied to the notion that the Canadian people make the final determinations with regard to their political realities. This is typically expressed through elections that are held at specified intervals and allow citizens to vote for the political policies, ideas, and values they most strongly endorse. We will examine the voting behaviour of Canadians more closely in Chapter 4 ("The Political Process"); however, here it is important that the concept of popular sovereignty be understood as a key component of our political culture. Elections are a mechanism of democracy that allows Canadians to regularly communicate their popular will. A special appeal to the people may also arise around some important issue that emerges between elections. This is when we can see referendums or plebiscites (e.g., the Quebec sovereignty referendum of 1999) that activate popular sovereignty in a specific manner. Canadians, however, are generally not inclined to this approach to decision-making.

Elections provide Canadian citizens with an opportunity to select those who will represent them within the federal Parliament (or in a provincial legislature) and to make decisions on their behalf. This approach is largely successful, although there is no shortage of thinking about ways in which elected members of Parliament may become more directly accountable to their constituents.

Political Equality   Another important aspect of our Canadian political culture is the concept of political equality. Political equality means that everyone who is eligible to vote is treated equally as a voter. "One person, one vote" is a hallmark of the most basic political equality, and this is now guaranteed in Canada. Some would argue, particularly from an elite, state-centred, or neo-Marxist perspective, that this political equality is actually a fiction. If political parties in Canada are extensively supported by elite corporate and economic interests, it seems likely that their attention will be toward those who have supported them financially rather than the humble voters. For example, Dyck (1993) states:

> Beyond election day itself, tremendous inequalities in political influence begin to emerge, such as in pressure group activity. Such nonelectoral inequalities, however, are beyond the scope of political equality as a bare ingredient of the definition of democracy. *(p. 166)*

Political Freedom   A third component of Canadian political culture is the idea of political freedom. In Canada, political freedom takes several forms. While there will be further discussion in Chapter 3 ("Canadian Constitutional Issues") with respect to the nature of our constitutional freedoms, they are essentially as follows:

- freedom of conscience and religion;

- freedom of thought, belief, opinion, and expression, including freedom of the press and other media of communication;

- freedom of peaceful assembly; and

- freedom of association.

Again, such freedoms are often seen as the symbolic anchors of any genuine democracy. Canada has a relatively strong tradition of observing these freedoms with only occasional lapses worthy of serious note. The prohibition of "unlawful associations" during the time of the Winnipeg General Strike in 1919, the incarceration of Canadians of Japanese extraction during the Second World War, and the imposition of the *War Measures Act* during the October Crisis in 1970 when the FLQ was threatening public safety in Quebec are examples that may challenge our claim to an absolute commitment to liberal democracy. Some observers have noted that Canadians have a particularly strong deference to authority that undermines their political freedoms.

Majority Rule   Majority rule holds that matters of important public policy are decided by the greatest number of supporters of one proposition over another. However, in most democracies, including Canada, majority rule immediately raises the question of minority rights. In our context, this can be seen in the treatment of various minority groups; for example, Roman Catholic and Protestant minority education rights were recognized in the *Constitution Act, 1867*. Another example would relate to the fact that minority language rights have been acknowledged in Canada.

What makes the Canadian political environment unique is Canada's beginning as New France and the impact this reality has had on our evolution as a sovereign entity. The "Dominion" of Canada represents a distinctive form of government and constitutes a particular historical entity with particular characteristics. There is indeed a "counter-revolutionary" quality to nation-building in the Canadian context and we have continuously seen ourselves as a more moderate and, yes, more cautious version of our neighbours to the south. Dyck (1993, pp. 169–177) has identified five basic Canadian values that help to distinguish us from the United States:

- balance between individualism and collectivism;

- particularism and tolerance;

- elitism, hierarchy, and deference to authority;

- egalitarianism; and

- caution, diffidence, dependence, and non-violence.

Basically, we appear to be more tolerant of difference in Canada and are not so quick to rely on the so-called rugged individualism that characterizes the United States. Dyck (1993) notes, for example:

> Related to elitism and hierarchy is another fundamental difference between Canadian and American values — the much greater deference to authority in Canadian society. Canadians demonstrate greater respect toward the law, judges, police, religious leaders, and many others with "legitimate power." Peace, order, and good government rather than individual liberty is the Canadian ideal, and many observers have noted that Canada is probably the only country where a police officer is a national symbol — certainly a contrast to the U.S. hero "Horatio Alger," the self-made man. *(p. 173)*

Canadians in general seem to be more trusting of their governments than Americans are and do not generally assume that their leaders are taking advantage of them. There is, by and large, a higher level of trust and respect for Canadian leaders, although it would seem this will not always remain so, in light of the battles fought over federal politics in the last few years in Ottawa. Certainly, citizens are beginning to be a little more reluctant to give their leaders the fullest measure of confidence. Many writers look back to the patterns of settlement in the Canadian West and compare it to the more chaotic circumstances found in the American West. Canadians have the example of the comparatively orderly North-West Mounted Police and the progressive imposition of law and order, while settlement of the western United States is remembered (if somewhat inaccurately) as a wild and restless process with more evidence of vigilantism than vigilance.

Canada is witness to the fairly successful implementation of what has been called multiculturalism, as opposed to the application of the "melting pot" theory that has been applied in the United States. There is substantial, and growing, evidence of the broad diversity of our Canadian population. On the negative side, however, we do have a particularly troubled history of neglect and confusion with regard to our treatment of the First Nations peoples in Canada. Their presence on the political landscape is undeniable, and yet our ability to develop and deliver justice and fairness to this important constituency has been singularly unimpressive.

Canadians continue to adhere to a fairly high degree of elitism, hierarchy, and deference to authority. We continue to respect authority, although much has happened in the last few decades to erode that reliable response, and we are slowly becoming more prone to questioning the voices of our corporate leaders, both public and private. In our domestic and international political affairs, Canada is a country that does appear to display a relatively high level of caution and diffidence. We have a strong dependence upon the United States that appears to wax and wane, depending on the government in power in Ottawa. Finally, we continue to be characterized by an adherence to non-violence, and our crime rate statistics tend to bear this out when compared with other nations, particularly the United States.

In summary, Dyck (1993) makes the following observations:

In short, Canada is a sprawling, decentralized country of great natural beauty and variety, an officially bilingual and multicultural society characterized by severe regional and ethnic strains, which fuses British parliamentary system with American federalism. It has a restrained commitment to democracy combined with deferential and elitist traits, and prefers a strong government. This reflects the strength of social democratic sentiments and has resulted in redistributive egalitarianism. Moreover, in an international reflection of its domestic character, Canada is a peaceful, constructive member of the world community. *(p. 178)*

Regional Distinctions

It is worthwhile to consider the whole question of regional disparity and what this has meant, and continues to mean, to Canada. It is a necessary feature of Canadian federalism that is often difficult to grapple with when these disparities become pronounced. The issues that define the regional disparities in Canada revolve around economic, ethnic, cultural, and linguistic distinctions in the following regions:

- Atlantic provinces (Newfoundland and Labrador, Prince Edward Island, Nova Scotia, and New Brunswick)

- Quebec

- Ontario

- Prairie provinces (Manitoba, Saskatchewan, and Alberta)

- British Columbia

- Northern Canada (Yukon, the Northwest Territories, and Nunavut).

Of course, even the process of identifying distinct "regions" predetermines some of the parameters for subsequent discussion of regional disparities. It should be expected that the fundamental nature of any given "region" will likely change over time. A serious consideration of any particular region will also be contingent upon the matter that is under review. For example, offshore fishing issues may be of serious concern to Newfoundland and Labrador but of little, or no, interest to the fisheries in Nova Scotia. Francophone issues may be a regional concern for people not only in Quebec, but also for significant portions of New Brunswick. Matters of Western "alienation" may influence voters from Manitoba to British Columbia. Finally, Northerners are not always some monolithic regional bloc for every concern and controversy that may motivate one portion of the North to political action.

Ethnic Distinctions

Canada's citizens can trace their ancestry back to virtually every nation of the world. We have evolved into a nation that can take pride in its strong representation of visible minorities. As a country that values multiculturalism, Canada has generally tended to respect the aspects of ethnic culture that distinguish various groups that have become part of our societal fabric. However, this is not to say that visible minorities have always found circumstances fully supportive and fair as they have sought out a new existence within the Canadian system. Table 2.1 outlines the details of our visible minority population (by age) based on 1996 census figures.

Table 2.1	Visible Minority Population by Age, 1996 Census						
	Total	0-14	15-24	25-44	45-64	65-74	75 and over
Total population	28,528,125	5,899,200	3,849,025	9,324,340	6,175,785	2,024,180	1,255,590
Total visible minority population	3,197,480	778,340	521,060	1,125,730	581,275	129,415	61,655
Black	573,860	170,870	96,895	186,995	94,520	16,025	8,555
South Asian	670,590	168,585	107,465	230,245	127,355	26,425	10,505
Chinese	860,150	171,110	135,580	299,815	177,980	50,680	24,990
Korean	64,840	12,115	15,525	19,475	14,610	1,765	1,340
Japanese	68,135	12,545	11,830	20,850	14,670	5,280	2,965
Southeast Asian	172,765	49,295	28,380	68,210	20,195	4,895	1,785
Filipino	234,195	50,985	33,995	90,100	45,370	8,845	4,900
Arab/West Asian	244,665	60,850	37,040	95,005	39,955	8,185	3,630
Latin American	176,975	46,530	31,575	68,500	25,190	3,670	1,500
Visible minority, n.i.e.[1]	69,745	15,065	11,015	27,590	12,995	2,160	915
Multiple visible minority[2]	61,575	20,385	11,755	18,945	8,425	1,480	575

Visible minority population: The *Employment Equity Act* defines the visible minority population as "persons, other than Aboriginal peoples, who are non-Caucasian in race or non-white in colour." The visible minority population includes the following groups: Chinese, South Asian, Black, Arab/West Asian, Filipino, Southeast Asian, Latin American, Japanese, Korean, and Pacific Islander.

1 Visible minority not included elsewhere. Includes respondents who reported a single write-in response indicating a Pacific Islander group (e.g., *Fijian or Polynesian*) or another single write-in response likely to be a visible minority group (e.g., *Guyanese, West Indian*).

2 Includes respondents who reported more than one visible minority group.

Source: Statistics Canada, 1996 Census *Nation* tables.

It is also useful to understand that Canada is a destination for immigrants from across the globe, and that different patterns of immigration have emerged over the last several decades. While prior to 1961, significant numbers of immigrants were coming to our country from the United Kingdom, Eastern Europe, and Southern Europe, today our newest citizens are arriving in substantial numbers from Central and South America; Southern, Southeast, and Eastern Asia; and Africa. Table 2.2 provides an outline of those statistics.

Class Distinctions

Canada may be viewed as being, in many ways, a classless society. However, there remain within the fabric of Canadian society a number of economic distinctions that clearly demonstrate that we have not been able to prevent certain portions of our society from falling well below the poverty line with respect to household income. The existence of food banks in virtually every major Canadian city is a fairly reliable indication that not everyone has reached a level of self-sufficiency that would appear to be consistent with our liberal democratic system.

Unlike the United Kingdom, which continues to have vestiges of a fairly rigid class system, Canada has no real upper class deriving privilege from birth and special entitlement. We have no members of our society who function as lords and ladies, barons and baronets. There is no landed gentry holding a special and exclusive title to honours and positions closed to the rest of Canadian society. We do, however, have those who are truly wealthy and incredibly powerful, people whose net worth is in the hundreds of millions of dollars. Unfortunately, we also have those who are deeply affected by poverty. These are the working poor, people who struggle with substandard wages and difficult or erratic working conditions. Most of our citizens occupy the middle class, what Karl Marx referred to as the "petite bourgeoisie." These are people who are gainfully employed and whose living standards are more than tolerable. They are typically well-educated, and they are in possession of a wide range of material goods, including homes, cars, and other modern conveniences.

Many of the struggles that define Canadian politics are in some manner derived from the tensions that develop among the various classes within society. Government policies and programs that are seen to unfairly advantage the wealthy are attacked by those who represent those struggling with poverty or the working poor. Members of the middle class frequently challenge governments to provide them with greater levels of financial support and to lighten the tax burden that diminishes their individual purchasing power. Of course, the wealthy are always seeking ways to protect and expand their power and influence through the political machinery of government. These battles are rooted in collective self-interest and are likely to remain in evidence in Canadian politics. We are far from the classless society envisaged by some political theorists.

Religious, Gender, and Age Distinctions

With regard to the religious divide across Canada, Statistics Canada is, again, able to provide some helpful figures pertaining to the representation of different faiths observed by Canadians. Table 2.3 provides a comparison of population by religion based on the 1981 and 1991 Canadian censuses.

Table 2.2   Immigrant Population by Place of Birth and Period of Immigration, 1996 Census, Canada

Place of birth	Total Immigrant Population	Period of Immigration				
		Before 1961	1961-1970	1971-1980	1981-1990	1991-1996[1]
Total—Place of birth	4,971,070	1,054,930	788,580	996,160	1,092,400	1,038,990
United States	244,695	45,050	50,200	74,015	46,405	29,025
Central and South America	273,820	6,370	17,410	67,470	106,230	76,335
Caribbean and Bermuda	279,405	8,390	45,270	96,025	72,405	57,315
United Kingdom	655,540	265,580	168,140	132,950	63,445	25,420
Other Northern and Western Europe	514,310	284,205	90,465	59,850	48,095	31,705
Eastern Europe	447,830	175,430	40,855	32,280	111,370	87,900
Southern Europe	714,380	228,145	244,380	131,620	57,785	52,455
Africa	229,300	4,945	25,685	58,150	64,265	76,260
West-Central Asia and the Middle East	210,850	4,975	15,165	30,980	77,685	82,050
Eastern Asia	589,420	20,555	38,865	104,940	172,715	252,340
South-East Asia	408,985	2,485	14,040	111,700	162,490	118,265
Southern Asia	353,515	4,565	28,875	80,755	99,270	140,055
Oceania and Other[2]	49,025	4,250	9,240	15,420	10,240	9,875

Immigrant population: Refers to people who are, or have been, landed immigrants in Canada. A landed immigrant is a person who has been granted the right to live in Canada permanently by immigration authorities. Some immigrants have resided in Canada for a number or years, while others are recent arrivals.

1 Includes only the first four months of 1996.

2 "Other" includes Greenland, St. Pierre and Miquelon, the category *other country*, as well as immigrants born in Canada.

Source: Statistics Canada, 1996 Census *Nation* tables.

Table 2.3	Population by Religion, 1981 and 1991 Censuses, Canada			
	1981		**1991**	
	Number	%	Number	%
Total Population	24,083,495	100.0	26,994,045	100.0
Catholic	11,402,605	47.3	12,335,255	45.7
Roman Catholic	11,210,385	46.5	12,203,620	45.2
Ukrainian Catholic	190,585	0.8	128,390	0.5
Other Catholic	1,630	—	3,235	—
Protestant	9,914,575	41.2	9,780,715	36.2
United Church	3,758,015	15.6	3,093,120	11.5
Anglican	2,436,375	10.1	2,188,110	8.1
Presbyterian	812,105	3.4	636,295	2.4
Lutheran	702,900	2.9	636,205	2.4
Baptist	696,845	2.9	663,360	2.5
Pentacostal	338,790	1.4	436,435	1.6
Other Protestant	1,169,545	4.9	2,127,190	7.9
Islam	98,165	0.4	253,260	0.9
Buddhist	51,955	0.2	163,415	0.6
Hindu	69,505	0.3	157,010	0.6
Sikh	67,715	0.3	147,440	0.5
Eastern Orthodox	361,565	1.5	387,395	1.4
Jewish	296,425	1.2	318,065	1.2
Para-religious groups	13,450	0.1	28,155	0.1
No religious affiliation	1,783,530	7.4	3,386,365	12.5
Other religions	24,015	0.1	36,970	0.1

— amount too small to be expressed

Source: Statistics Canada, Catalogue no. 93-319-XPB.

There is a great deal of statistical information available with regard to the gender divisions across Canada. It is, for instance, interesting to note that women outnumbered men in all of the provinces. This is especially the case in Atlantic Canada. In Nova Scotia in 2001, there were 468,920 women and 439,090 men, which represents a ratio of 93.6 men for every 100 women. In the territories, however, men outnumber women. The highest ratio is found in Nunavut (Canada's newest territory) with 107.2 men for every 100 women.

In Canada's older population, there appears to be a trend of senior men gaining on senior women. Mortality rates in the past meant that women typically outlived men and, therefore, made up a significantly higher percentage of the elderly population. However, senior men appear to be living longer. In 2001 there were 75 senior males for every 100 senior women. This represents an increase from 72 senior males in 1991. Prior to 1951 senior males outnumbered senior females, in large part due to the regularity of women dying in childbirth.

Demographics are important to our understanding of the political realm. These numbers tell us what we can expect in terms of population growth and population change. Significant shifts in various categories will have an impact on the services that will be needed by Canadians and will directly influence government policies and programs. With regard to the age of our population, it has been established that the median age has reached an all-time high. Based on recent census data, the median age of Canada's population is 37.6 years. This represents an increase of 2.3 years from 1996 and is the largest census-to-census increase in 100 years. This increase is significant on several fronts. It results from a major decline in the number of births that have occurred since 1991. This has contributed to a continuous rise in the median age since the end of the post-war baby boom in 1966, when it was calculated at 25.4 years.

Canada's aging population has important implications for the economy, the labour force, and the delivery of social services, including a predictably high demand on the nation's health care system. In 2001 the number of Canadians aged 80 and over reached 932,000. By the year 2011, there will be more than 1.3 million persons 80 years of age or older. Clearly, this will place an enormous strain on our country's resources and services.

Federalism

Canadian politics is characterized by a federal system of government. This structure entails a central government with several federated entities affiliated in a formal arrangement that allows them to share resources and exercise responsibilities. Over time, the nature of Canadian federalism has evolved considerably (Meekison, 1977; Smiley, 1972; Smiley, 1987). There has been a trend toward more autonomous provincial governments and a movement away from a totally dominant federal government in Ottawa. The *Constitution Act, 1867,* provides the outline of the division of powers between the central government and the provinces. Section 91 enumerates the exclusively federal powers, while section 92 sets out those of the provincial governments.

Many battles have been fought in attempting to clarify, or modify, the specific application of the rules of Canadian federalism. Increasingly, the provinces are demanding more control of their resources and are lobbying the federal government for increased transfer payments in order that they may introduce and sustain their own particular programs of social and economic development. The issue of federalism is critically important to the shaping of Canadian identity. There have been many examples of compromise and accommodation, but also many instances of conflict and acrimony. In our examination of specific constitutional controversies in Chapter 3 ("Canadian Constitutional Issues"), we will learn more about the nature of Canadian federalism.

THE CORE INSTITUTIONS

The Canadian Parliament is made up of three components:

- the Crown;
- the Upper House, or Senate; and
- the Lower House, or House of Commons.

The Crown

As the Sovereign of Canada, the Queen holds the executive authority under section 9 of the Constitution. The head of government is, indeed, the Prime Minister; however, the Governor General is the one who carries out the official functions of the head of state. The Governor General technically has no political affiliation and is appointed by the Queen, on the advice of the Prime Minister, as her personal representative in Canada.

Among the official duties of the Governor General are the following:

Choosing the Prime Minister of Canada. Convention now sees the Governor General selecting the leader of the party with the majority of seats in the House of Commons. If no party holds a majority, the Governor General may call upon the party leader who is most likely to maintain the confidence of the House of Commons as the Prime Minister.

Summoning Parliament. The Governor General officially convokes Parliament, provides royal assent to legislation, and signs state documents.

Dissolving Parliament. Under current convention, this means that the Prime Minister will provide advice to the Governor General moving toward an election within five years of a Government's tenure. Should the Government be defeated in a vote of non-confidence, the Governor General may decide in the best interests of Canadians, and with the advice of the Prime Minister, to proceed with a general election. Alternatively, the Governor General may request that the official Opposition form the next Government.

Also, the Governor General plays an important symbolic role within the Government of Canada. This person, who is the formal representative of the Queen, performs many ceremonial duties such as recognizing the significant achievements of individual Canadians; receiving delegations, diplomats, and dignitaries from foreign countries; and officiating at many other ceremonial events.

The Canadian Senate

The Fathers of Confederation hoped that the Upper House, or Senate, would provide the Government of Canada with a body of individuals who would provide sober second thought to the legislation that was being proposed through the Lower House. The Senate would be the place where careful consideration would ensure that bills that were not completely in the public interest would be turned back to the House of Commons. Furthermore, the Senate was intended to provide Canadians with a mechanism for safeguarding specific minority, provincial, or regional interests.

The 105 members of the Senate are summoned to office by the Governor General, on the advice of the Prime Minister. These individuals have affiliations with many different political parties, or may serve as independent members. The seats within the Senate are apportioned in a manner that provides equal representation for each region in Canada. In order to serve as a Senator, a person must be over 30 years of age, must own property, and must reside within the region represented. Senators take part in the debates held within the Senate Chamber, they review legislation and government estimates, and they investigate policy matters and other issues pertinent to Canadian citizens. The Senate functions along committee lines, similar to those maintained within the House of Commons. Bills may be introduced in the Senate, as long as they do not raise or allocate public funds. To receive royal assent and become Canadian law, a bill must be passed by both the Upper House and the House of Commons. Senators may serve the public until the age of 75.

The House of Commons

This is the national assembly that serves to represent the Canadian public. Currently, there are 301 seats in the House of Commons. These seats are distributed among the provinces in accordance with the population. No province can have fewer seats in the House of Commons than it has members in the Upper House.

The Government must maintain the support of the House of Commons if it is to continue to function in that role. This support is referred to as the confidence of the House. Should the Government lose a vote dealing with a major policy issue, it is expected that the Government will resign and seek the Governor General's approval for a general election. This may include a Government budget or tax bill. This practice of sustaining the confidence of the House of Commons is a feature of representative government and demonstrates clearly that the executive branch (i.e., the Prime Minister and Cabinet) cannot continue to function without the consent of the members who are elected to the House of Commons. This constitutes one element of public accountability.

Members of Parliament are elected to the House of Commons to represent Canadian voters within the ridings, or constituencies. Candidates with a plurality of votes are awarded the seat for any given riding or constituency. These individuals hold their seats for the life of the Parliament (i.e., for a maximum of five years). Members are entitled to seek re-election.

Within the House of Commons there are two main divisions: Government and Opposition. All members of Parliament who are not affiliated with the Government party are part of the Opposition. In recent years there has been considerable discussion (Docherty, 2000) around the challenges and technicalities of dealing with a five-party House of Commons. Following the 1993 general election there were five officially recognized official parties due to the fact that the Bloc Québécois, the Liberals, the New Democratic Party, the Progressive Conservatives, and the Reform Party each had the minimum of 12 seats in the House of Commons. We will deal with the aspects of party politics, including this new reality of the Canadian Parliament, in Chapter 4.

The members of Parliament in the House of Commons occupy themselves with the routine, and extraordinary, business of government. They formulate and debate legislative matters that may take the form of laws. They serve on committees to carefully consider policy matters of importance to Canadians and report their findings to the public through Parliament. Furthermore, Members maintain contact with their local constituents and work to maintain their party's position within the House of Commons.

The Prime Minister

The Prime Minister is, in essence, the chief executive officer of the country. The Prime Minister selects the members of the Cabinet and is responsible for setting the agenda for Cabinet meetings which in turn drive the agenda for the country. As Dyck (1993) notes:

> The system of government that Canada inherited from Britain has traditionally
> been called "cabinet government," but most observers argue that such a label does

not do justice to the modern pre-eminence of the prime minister. Political scientists and practitioners disagree as to whether cabinet government has been completely transformed into a system of "prime ministerial government," but no one doubts that the prime minister has enormous power and should be singled out for special attention. *(p. 416)*

In many instances, Prime Ministers have placed a particular stamp on the nature of their government and the country. John A. Macdonald, Wilfrid Laurier, Lester B. Pearson, John Diefenbaker, Pierre Trudeau, Brian Mulroney, and Jean Chrétien have all added depth and dimension to the role of the Prime Minister in Canada.

The Prime Minister is also the leader of the party in power and therefore the focus of political life in Canada. The Prime Minister is the country's chief policy-maker. Prime Ministers have a major role in the House of Commons and are stars of the televised debates that take place in our federal government. The Prime Minister is responsible for the appointment of judges, Senators, deputy ministers, and heads of a wide range of government agencies, boards, and commissions. The Prime Minister is often the country's chief diplomat and represents Canada abroad. This is particularly the case in a climate of globalization where international meetings are the norm rather than the exception. If one considers the G-8 summit at Kananaskis, and other global meetings that reflect the increasing interconnection among nations, it becomes obvious that the role of the Prime Minister is a pivotal one. The Prime Minister is Canada's leading actor on the world stage.

The Cabinet

The Cabinet is the decision-making body *par excellence* for many areas of government policy. As Doern and Aucoin (1979, p. 27) indicate, the Cabinet, in conjunction with the Prime Minister and the central agencies, "constitute the heart of the executive-bureaucratic arena of the Canadian political system and hence of the day-to-day policy processes of government." This section will focus on

- the Solicitor General and the Minister of Justice (and Attorney General of Canada) as two Cabinet members who have an enormous impact on the justice system in Canada;

- the importance of having some kind of regional or provincial representation in the Cabinet and the need to acknowledge the talents of veteran members of the party in power;

- the role of the Deputy Prime Minister and some of the key Cabinet posts;

- francophone and other representative issues including the need for women in the Cabinet;

- the concept of Cabinet solidarity and its importance for decision-making at this level;

- Cabinet secrecy and the issue of confidentiality as issues are being discussed; and

- protecting state secrets and the inner workings of government.

The Prime Minister is the leader of the Cabinet and makes decisions about the size, membership, and responsibilities of that body. The Prime Minister's "team" or ministry consists of both ministers and secretaries of state. While secretaries of state are sworn members of the Privy Council, they are not members of the Cabinet. These individuals are allotted important responsibilities in support of the work of the Cabinet ministers to whom they are assigned. They are entrusted with duties in areas that the Prime Minister has identified as key government priorities. What follows is a listing of existing Cabinet portfolios and state secretariats.

Cabinet Portfolios:

- Labour
- National Revenue
- Environment
- Justice and Attorney General of Canada
- Citizenship and Immigration
- Transportation (with responsibility for Crown Corporations: Canada Mortgage and Housing Corporation, Canada Post Corporation, Royal Canadian Mint, Canada Lands Company Limited, and the Queen's Quay West Land Corporation)
- Canadian Heritage
- Natural Resources
- Intergovernmental Affairs
- Public Works and Government Services
- Foreign Affairs
- Solicitor General of Canada
- Finance
- National Defence
- Health
- Indian Affairs and Northern Development
- Veteran Affairs
- International Trade
- Industry
- Human Resources Development
- Fisheries and Oceans
- Agriculture and Agri-Food
- International Cooperation

State Secretariats:

- Multiculturalism
- Status of Women
- International Financial Institutions
- Children and Youth

- Atlantic Canada Opportunities Agency

- Amateur Sport

- Economic Development Agency of Canada for the Regions of Quebec

- Asia-Pacific

- Central and Eastern Europe and Middle East

- Rural Development

- Federal Economic Development Initiative for Northern Ontario

- Western Economic Diversification

- Indian Affairs and Northern Development

- Science, Research and Development

- Latin America and Africa

- Francophonie

Cabinet members meet regularly to discuss important issues that concern the business of government and which are vital to the interests of Canadians. Among the broad topical areas that the Cabinet addresses on a routine basis are

- major government spending decisions;

- proposals on new or amended laws for Parliament to debate; and

- major new programs, services, or policies for the federal government.

Due to the intensity and complexity of the work of the Cabinet, there are a number of smaller Cabinet committees in place that allow selected ministers to undertake preliminary discussions on relevant public policy issues within their scope of responsibility. Currently, the Cabinet operates through the five formal committees listed below.

Economic Union.   Examines issues with potentially significant economic impacts such as industrial or financial matters, natural resources and agricultural concerns, and other related issues.

Social Union.   Examines proposals from various departments that deal with initiatives aimed at supporting the growth of Canadian society. Matters that this committee has dealt with in the past include health, childhood learning, and safe and secure communities.

Treasury Board.   Has an overall focus on government management and includes measures to ensure the cost-effective use of public funds.

Special Committee of Council.   Has primary responsibility for the review and approval of new and amended regulations which are developed to support existing Canadian legislation.

Government Communications.   Examines government-wide communications issues in order to ensure that the Canadian public is well informed on government priorities and that public feedback is processed in a consistent manner.

The Solicitor General of Canada

Overall, the Department of the Solicitor General of Canada, which was created in 1966, is dedicated to protecting Canadians and maintaining a safe and peaceful society within Canada. Specifically, the Solicitor General is in charge of the following areas of public service activity:

- Royal Canadian Mounted Police (RCMP)

- Correctional Service of Canada

- Canadian Security Intelligence Service (CSIS)

- National Parole Board

Within this broad portfolio, the Solicitor General has much to do with public safety and security. The Solicitor General is responsible for a departmental budget of more than $3.2 billion and holds overall accountability for a staff of 38,000 public servants.

The portfolio of the Department of the Solicitor General is reflected in Figure 2.1.

Of particular importance to the law enforcement area is the Solicitor General's responsibility for the RCMP. This means that the Solicitor General is in charge, in a broad, strategic manner, for the functioning of the nation's police service. Furthermore, the Solicitor General holds overall accountability for the RCMP in its capacity as both a provincial and a municipal police service in several jurisdictions across Canada. Accordingly, this Cabinet position is extremely important when one considers the extensive mandate the Solicitor General holds.

As a federal government department, the Solicitor General's office provides significant levels of support to the policing and law enforcement community. It is always useful to review some of the publications that have issued from this department in order to keep up to date on developments in law enforcement, policing, and public safety. What

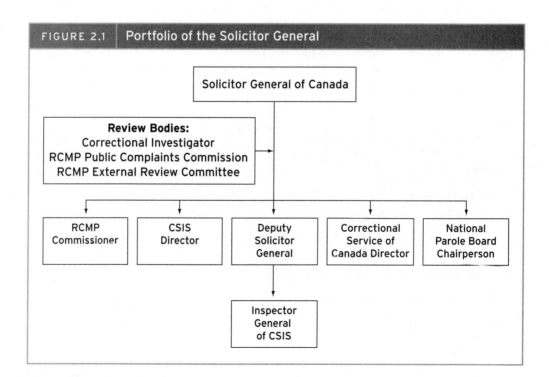

FIGURE 2.1 | Portfolio of the Solicitor General

follows is a selection of topics dealt with by various publications distributed under the auspices of the Solicitor General of Canada:

- legislation governing a DNA data bank;

- cross-border crime issues;

- proposed legislation for the prevention and control of terrorism;

- approaches to combating transnational crime;

- use of electronic surveillance;

- factors affecting police diversion of young offenders;

- national forum on youth gangs;

- organized crime impact studies; and

- exemplary school-based police programs to combat youth violence.

The publications provide a good representation of the research interests of the Solicitor General's department, and they help to support the activities of individual police services across Canada, including the RCMP.

One pre-eminent area of activity is the work of the Solicitor General's department in the realm of Aboriginal policing. This department holds responsibility for the process by which Aboriginal self-policing will be realized in Canada.

It is worthwhile to examine this federal government department in some considerable detail in order to gain some deeper appreciation of the complexity of the Canadian government structure. This examination is appropriate because it reveals the depth of involvement that is seen in virtually every government department. It will also allow us to come to a better understanding of the specific engagements that are required for a department that is dedicated to the policy, program, and priority issues relevant to law enforcement and policing in Canada.

The office of the Deputy Solicitor General is where the day-to-day functioning of the department is guided. The Deputy Solicitor General is the senior civil servant within the department and is the link between the bureaucracy and the elected level, here in the person of the Solicitor General. It is instructive to consider the layout of this portion of the organization. Figure 2.2 shows the organizational structure of the Deputy Solicitor General's area of responsibility.

With the Deputy Solicitor General as its corporate leader, this department carries out all of the roles and responsibilities that are necessary to support the Solicitor General in his or her legislated duties. At the functional level, the department is structured as follows:

- National Security Branch

- Strategic Policy and Programs Branch

- Policing and Law Enforcement Branch

- Corporate Management Branch

- Legal Services Branch

- Office of the Inspector General, CSIS

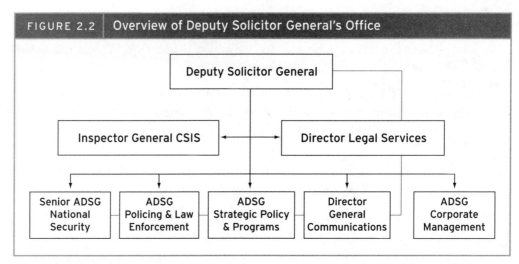

FIGURE 2.2 | Overview of Deputy Solicitor General's Office

Note: ADSG = Assistant Deputy Solicitor General

It is noted that there is an important communications function that is included in the organizational chart. This discussion will focus particularly on the work of the Policing and Law Enforcement Branch.

The Policing and Law Enforcement Branch serves to provide advice on a broad range of issues that influence the role of the Solicitor General in the policing and law enforcement policy field. It also operates to support the Solicitor General's role as the Cabinet minister responsible for the RCMP. The branch provides the minister with independent advice in order to manage this government portfolio. It also functions to formulate policy guidance on sensitive issues and assists the minister in the conduct of department business as it pertains to other government departments, the Cabinet, Parliament, and provincial and municipal levels of government, as well as the policing community and the public.

This branch is divided into three areas of responsibility:

- Law Enforcement Division
- Organized Crime Policy and Coordination Division
- Policing Policy Division

Law Enforcement Division   The Law Enforcement Division is responsible for

- developing, steering, and implementing legislative initiatives aimed at enhancing law enforcement (e.g., amendments to the DNA data bank legislation, police powers for organized crime, drug enforcement, and drug control regulations);
- initiating, developing, implementing, and evaluating national law enforcement policies, programs, and strategies aimed at addressing identified weaknesses in the current law enforcement system;
- supporting the police role in community-based partnerships that enhance safety and prevent crime and victimization;
- leading evaluations of specified horizontal multi-departmental initiatives aimed at combating organized crime (e.g., Anti-Smuggling Initiative, Integrated Proceeds of Crime);

- supporting the department's role in law enforcement policy and related issues by participating in, and supporting key international forums (e.g., the Inter-American Drug Abuse Control Commission and the World Health Organization Framework Convention on Tobacco Control); and

- managing a program of contributions to support a range of related interests (e.g., associations, conferences, workshops, and publications related to crime prevention, law enforcement, and illicit drug control).

Organized Crime Policy and Coordination Division   The Organized Crime Policy and Coordination Division is responsible for

- developing policy proposals and contributing toward the development of national and international strategies and initiatives against organized crime;

- providing research, logistical, and policy analysis support to the department and the national strategy on organized crime;

- supporting the department's chairmanship of the National Coordinating Committee on Organized Crime, the Federal-Provincial-Territorial Deputy Ministers' Steering Committee on Organized Crime, and the Canada-US Cross-Border Crime Forum; and

- participating in a variety of international forums on transnational organized crime, including the G-8, the Financial Action Task Force, and the United Nations.

Policing Policy Division   The Policing Policy Division is responsible for

- providing ongoing liaison, policy advice, and coordination with the RCMP and the eight provinces and three territories that have entered contracts for RCMP provincial or territorial policing;

- developing and providing advice to the minister with respect to the minister's civilian oversight responsibility for the RCMP, including the corporate, administrative, and operational roles and policies of the RCMP, labour relations and accountability issues, and advice to the minister on the roles and recommendations of the RCMP External Review Committee and the Commission for Public Complaints Against the RCMP;

- providing ongoing liaison and preparing of advice for the minister on policy proposals with implications for the RCMP, including preparation or review of Cabinet documents and Treasury Board submissions; and

- providing policy development, evaluation, and operations with respect to the government's Framework Security Policy for provincial and municipal costs of Prime Minister-led summits and minister-led meetings.

Department of the Solicitor General of Canada Statutes:

- *Canadian Security Intelligence Act*

- *Charities Registration (Security Information) Act*

- *Corrections and Conditional Release Act*

- *Criminal Records Act*

- *Department of the Solicitor General Act*

- *DNA Identification Act*

- *Prisons and Reformatories Act*

- *Royal Canadian Mounted Police Act*

- *Royal Canadian Mounted Police Pension Continuation Act*

- *Royal Canadian Mounted Police Superannuation Act*

- *Transfer of Offenders Act*

- *Witness Protection Act*

A Few Words Concerning Communications and Media Relations   The high profile of certain activities undertaken by law enforcement and policing agencies (e.g., the pepper spray incident at the APEC meeting in Vancouver in 1998, where RCMP officers were accused of using excessive force against peaceful protestors), makes it appropriate to spend a little time considering the communications issues that must be dealt with as part of the bureaucratic level of government. As noted above, much of the work done within the department is aimed at informing, advising, and protecting the minister. To that end, a significant amount of time and energy is devoted to the communications function within this department. This is true in every government portfolio; however, policing is one policy area where careful management of corporate communications is absolutely critical. Within the Department of the Solicitor General of Canada, reporting to the Deputy Solicitor General, the Communications Group provides the following services:

Media monitoring.   Members of the department are connected through their desktop computers to an electronic media monitoring service. This service is continuously updated throughout the day with relevant stories and provides fairly comprehensive coverage of all major English and French newspapers. This allows members of the department to be aware of late-breaking stories or high-profile items that may affect the organization.

Media analysis.   Members of the Communications Group prepare media analyses on an as-needed basis. They monitor media coverage following a major announcement by the minister, or they assess media coverage of issues of interest and importance to the minister. The Communications Group also contributes to the department's overall environmental scanning process which plays a role in the organization's strategic planning.

Media relations.   Officers within the Communications Group are routinely in contact with members of the media and often serve as backup spokespersons for the department in the absence of the minister's press secretary. Staff also refer telephone calls on behalf of the minister's office and coordinate the public response to inquiries from the media with departmental agencies and other federal departments.

News releases, speeches, and various writing tasks.   Members of the Communications Group are regularly engaged in the preparation, production, translation, and dissemination of news releases and speeches for the minister and the Parliamentary

Secretary. Group members work in consultation with senior department officials prior to submitting items to the minister's office for approval.

Communications advice and planning.   The department assists with planning and coordination of major announcements. Members of the Communications Group provide advice and guidance on the most effective means of communicating departmental announcements and articulate the substance of anticipated public reaction to these announcements. The group also provides objective commentary on communications plans or strategies prepared by agencies within the department.

Coordination.   The Communications Group coordinates departmental announcements and related media relations activities. This may involve other federal departments, as well as several agencies within the department. Members of the group are called upon to work closely with the Communications and Consultation Secretariat of the Privy Council Office and indirectly, with the Prime Minister's Press Office.

Publications.   The department publishes a wide range of reports and documents. Members of the Communications Group are frequently engaged in the editing, design, printing, and dissemination of these publications.

Internet.   The department maintains a home page, which was launched in March 1996. Members of the Communications Group ensure that all ministerial news releases and speeches are available on the site. They are also responsible for maintaining current biographical information on the minister, and they provide listings for published reports and other relevant departmental information.

Department of Justice Canada
(and Attorney General of Canada)

The federal Minister of Justice carries a portfolio that is, in many ways, closely linked to that of the federal Solicitor General. Both ministers deal with elements of the justice system and the Minister of Justice holds responsibility and accountability for the following areas of activity:

- crime prevention
- international law
- victims of crime
- criminal law
- Aboriginal justice
- e-government
- human rights
- dispute resolution
- youth justice
- family law
- biotechnology
- intellectual property

In summary, the mission of the Department of Justice Canada is as follows:

- support the Minister of Justice (and the Attorney General of Canada) to ensure that Canada is a just and law-abiding society with an accessible, efficient, and fair system of justice;

- provide high-quality legal services and counsel to the Government of Canada and to client departments and agencies; and

- promote respect for rights and freedoms, the law, and the Constitution.

Essentially, the Department of Justice seeks to provide Canadians with a sound and reliable justice system. In that capacity, it assists other federal departments to develop policies and new laws that will achieve that purpose. It is also charged with responsibility for guiding the reform of existing federal laws. The Department of Justice serves as the Government's "lawyer" and, therefore, it provides legal advice, prosecutes cases under federal laws, and represents the Government of Canada in court. The Minister of Justice is simultaneously the Attorney General of Canada, which means that he or she is the chief law officer of the Crown in Canada.

Unlike other federal government departments, nearly half of the Department of Justice staff members are lawyers. The rest of the department's staff members bring various types of expertise in areas like research, the social sciences, and communications. There are 13 regional offices of the department across Canada, and the staff members in those offices provide legal advice to a range of federal departments and agencies. Staff members undertake a substantial amount of the Government's litigation within the provinces.

The Department of Justice is responsible for providing legal advice to the Government of Canada. In that capacity, staff members work to develop, reform, and provide interpretation for the country's laws. The department considers all proposed legislation and determines whether it may run afoul of the pre-eminent constitutional legislation: the *Canadian Charter of Rights and Freedoms*. Furthermore, members of staff provide direction on whether new legislation is valid from a legal perspective. They also ensure that legislative drafting is clear in both official languages and is consistent with the civil law system in place in the province of Quebec and the common law tradition in existence in the other provinces.

In the realm of policy-making, the Department of Justice is quite active. This is a result of the fact that many of the department's policies will have a direct impact on the lives of Canadian citizens. Issues such as fairness, access to justice, and fundamental rights are dealt with by the Department of Justice.

The department has an important role to play with regard to public safety and security. This is profoundly true since the events of September 11, 2001, in the United States and the attached threat of international terrorism. The *Anti-Terrorism Act* that came into force on December 24, 2001, was formulated within the Department of Justice and, although still controversial, was intended to provide appropriate authorities with the wherewithal to protect Canadians from terrorism. Also, the Department of Justice has been extensively involved in amending the *Criminal Code* to provide higher levels of authority for police agencies and prosecutors to deal with criminal organizations.

In the area of youth justice, the department has responded to a call from Canadians for careful attention to the system of youth justice. There has been growing concern about the efficacy of the *Young Offenders Act* and Parliament has responded with new legislation, the *Youth Criminal Justice Act*. The Department of Justice launched a Youth

Justice Renewal Initiative that, along with the new legislation, has three main objectives: ensuring that young offenders are compelled to face meaningful consequences for their actions; introducing improvements in the efforts at rehabilitating young offenders; and placing much greater emphasis on the prevention of youth crime.

In a broader context, the department provides support for many efforts at crime prevention — support that is applauded by police services across the country. The expectation is that by attacking the root causes of crime, such efforts can reduce crime and disorder problems substantially. The Department of Justice is the corporate sponsor for the National Strategy on Community Safety and Crime Prevention. This strategy supports a social development approach to crime prevention and seeks to address the causes of crime that may be found in the family, in society, in schools, or in other areas that can be dealt with in a proactive manner. The strategy places a special emphasis on youth and children, Aboriginal peoples, and women, as well as seniors, persons with disabilities, and ethno-cultural minorities.

Gun Control   The Department of Justice has done a great deal of work in the area of gun control in Canada. It was largely responsible for the development of the *Firearms Act,* which requires the licensing of all firearms owners and the registration of all firearms. The department works with provincial and territorial authorities through the Canadian Firearms Centre to coordinate and consolidate the implementation of the *Firearms Act.*

International, Federal, and Provincial Responsibilities   Because Canada is constructed along the lines of a federal system of government, law-making authority is shared between the Government of Canada and the provincial jurisdictions. The federal government bears responsibility for those areas of legislation that affect all Canadians, including criminal law, telecommunications, immigration, and other cross-jurisdictional matters. The provinces have responsibility for areas like education, health, and property.

While the majority of *Criminal Code* offences are prosecuted by Crown counsel within the provinces, the Department of Justice (acting on behalf of the Attorney General of Canada) prosecutes all other federal laws, including drug offences. The department is also responsible for the fulfillment of Canada's obligations under various international treaties, including extradition.

It is useful to note that the Department of Justice provides various forms of grants and assistance to support research into innovative approaches that may benefit the justice system in Canada. Projects considered for funding seek to contribute to a safer society, promoting access to justice, equality, and human rights. The department has made grants to projects that study family violence, victims of crime, youth justice, community safety, and approaches to crime prevention. As with the federal Solicitor General's department, Justice Canada is responsible for a considerable number of publications that are relevant to those interested in policing, public safety, and criminal justice. The following listing is only a sample of the titles that have been issued under the auspices of the Department of Justice:

- *Barriers to justice: ethnocultural minority women and domestic violence — a preliminary discussion paper*

- *Charging and prosecution policies in cases of spousal assault: a synthesis of research, academic, and judicial responses*

- *Disproportionate harm: hate crime in Canada — an analysis of recent statistics*

- *Educational programs that alter knowledge, attitudes and behaviour of youth: report no. 1*

- *Electronic money laundering: an environmental scan*

- *Firearms homicide, robbery and suicide incidents investigated by the Winnipeg Police Service (1995)*

- *Incarcerated gang members in British Columbia: a preliminary study*

- *Inuit women and the Nunavut justice system*

- *Public perception of crime and justice in Canada: a review of opinion polls*

- *Review of multiculturalism and justice issues: a framework for addressing reform*

- *A statistical profile of female young offenders*

- *Students' knowledge and perceptions of the* Young Offenders Act: *final report, Ontario Institute for Studies in Education, report no. 4*

- *Survey of sexual assault survivors*

- *Women speak: the value of community-based research on woman abuse*

Bureaucracy

In Canada, the bureaucracy plays an incredibly important role in making the rules for the day-to-day management of public affairs. At the federal level, the Public Service Commission of Canada provides several layers of administrative support and subject matter expertise. It is useful to recall the discussion about the theories and principles articulated by Max Weber in Chapter 1 and the importance of appreciating the pivotal role of an objective bureaucracy for effective public administration.

The bureaucracy represents the real machinery of government. The large number of departments within the federal government today and the requirements of a complex modern society place high demands on the bureaucracy to administer and monitor the processes, policies, programs, and procedures of government.

The mission of the Public Service Commission of Canada is to

- maintain and preserve a highly competent and qualified public service in which appointments are based on merit; and

- ensure that the public service is non-partisan and its members are representative of Canadian society.

To develop a stronger staffing system, the Public Service Commission of Canada has formulated a strategy that identifies several result and process values, as well as certain management principles that will guide its approach to staffing over the next several years. Figure 2.3 captures the key elements of that strategy.

Furthermore, to guide the actual work of civil servants as they operate on behalf of Canadians, a number of key functional values have been articulated by the Public Service Commission of Canada. These are important to everyone who serves the public interest.

FIGURE 2.3	Public Service Strategy

Results
- Representativeness: a public service that reflects the labour market
- Non-partisanship: staffing that is free of political or bureaucratic patronage
- Competence: public servants qualified to do their jobs

Process
- Fairness: fair treatment of employees and applicants
- Equity: equal access to employment opportunities
- Transparency: open communication about staffing practices and decisions

Management
- Flexibility: staffing that is adapted to organization's needs
- Affordability/Efficiency: simple, timely and effective staffing

Source: Public Service Commission of Canada. Used with permission.

Public Service Values:

Merit.   Recruitment and promotion should be based on skills and qualifications appropriate to the job, not on favouritism or political affiliation.

Political neutrality.   Public servants must serve the government of the day and speak truth to power. Public servants are free to vote but should not bring their political beliefs to the workplace.

Accountability.   Originally, this value was interpreted as administrative accountability, concern for the way in which things are done, but increasingly it is focused on the outcome of actions.

Anonymity.   This convention is closely bound to political neutrality and ministerial responsibility. Politicians protect the anonymity of public servants in exchange for their neutral advice. Public servants cannot speak in the House and so rely on ministers to defend their actions. This convention is under threat for different reasons; for example, public servants are increasingly in the public eye through public consultations and are encouraged to speak to media. There have been cases where public servants were publicly named by their ministers without an opportunity to clear their reputations (Al-Mashat).

Economy, efficiency, and effectiveness.   These three watchwords for the conduct of public servants tie into the concept of public servants being the stewards of taxpayers' money and guardians of the public's interests.

Responsiveness.   Public servants are expected to be sensitive to political and public needs.

Representativeness.   The public service should reflect the composition of the population at large, in terms of gender, religious, ethnic, and socioeconomic groups.

Fairness and equity.   Public servants should deliver their service to citizens without special favour.

Integrity.   Closely tied to honesty and judgement, integrity is fundamental to nurturing public trust in government. Conflict of interest guidelines are one means of encouraging integrity. (Source: Institute on Governance. Used with permission. See the Public Service Commission of Canada website, **http://www.edu.psc-cfp.gc.ca/ tdc/learn-apprend/psw/hgw/valeurs_e.htm**)

For a more detailed discussion of the bureaucracy, see Chapter 8 ("Canadian Government Bureaucracy").

Judiciary

The judicial branch is the third important cornerstone in the edifice of the Canadian parliamentary system, along with the legislative and executive branches. It is critical that we enjoy an independent judiciary and that the rule of law is a pre-eminent value within our society. The basic structure of the judicial branch is outlined in Figure 2.4.

Since 1949, with the abolishment of final appeals to the Privy Council in Britain, Canada's highest court has been the Supreme Court of Canada. The Supreme Court of Canada renders careful decisions on those cases that come before it and provides a final court of appeal for legal matters, including the interpretation of the Constitution. The appeal process currently applied within the Supreme Court of Canada is presented in Figure 2.5.

The Supreme Court of Canada consists of nine judges, three of whom originate from the province of Quebec. These judicial officers are appointed by the Governor General, upon the recommendation of the Prime Minister, and are able to hold office until they reach the age of 75.

The organization of the court system in Canada comprises four tiers. This four-tier structure is represented in Figure 2.6.

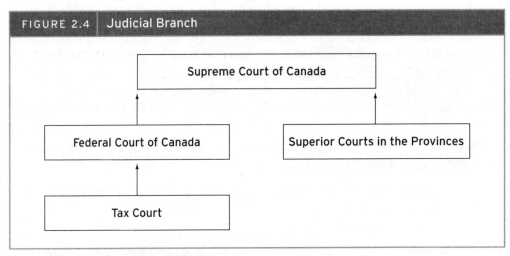

FIGURE 2.4   Judicial Branch

Source: Institute on Governance. Used with permission.

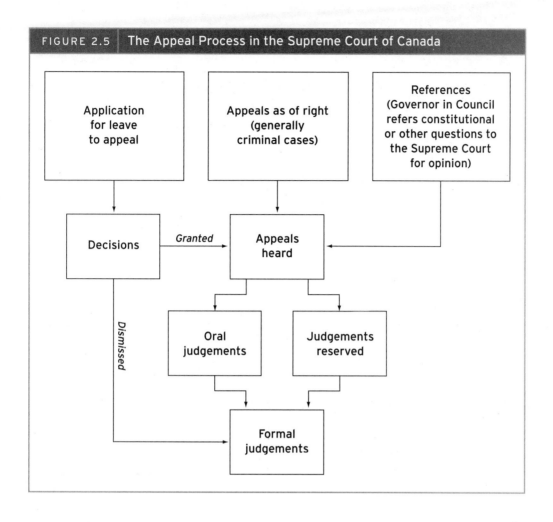

FIGURE 2.5 | The Appeal Process in the Supreme Court of Canada

Provincial and Territorial Government Structures

As noted earlier, within the federalist system of government, the Government of Canada has a special role that is defined with a certain degree of clarity in section 91 of the *Constitution Act, 1867*. Similarly, the nature of our 10 provincial governments is articulated, in general terms, in section 92 of that same piece of legislation. Accordingly, provincial governments have exclusive authority to pass laws that have an impact upon the following areas:

- direct taxation within the province for raising revenue for provincial purposes;
- borrowing of money on the sole credit of the province;
- establishment and tenure of provincial offices and appointment and payment of provincial officers;
- management and sale of public lands belonging to the province, including the timber and wood found on those lands;
- establishment, maintenance, and management of public and reformatory prisons in and for the province;
- establishment, maintenance, and management of hospitals, asylums, charities, and charitable institutions in and for the province (excluding marine hospitals);

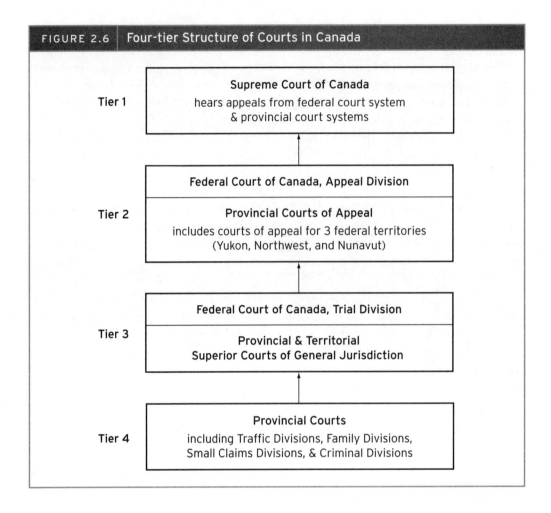

FIGURE 2.6 | Four-tier Structure of Courts in Canada

Tier 1

Supreme Court of Canada
hears appeals from federal court system
& provincial court systems

Tier 2

Federal Court of Canada, Appeal Division

Provincial Courts of Appeal
includes courts of appeal for 3 federal territories
(Yukon, Northwest, and Nunavut)

Tier 3

Federal Court of Canada, Trial Division

**Provincial & Territorial
Superior Courts of General Jurisdiction**

Tier 4

Provincial Courts
including Traffic Divisions, Family Divisions,
Small Claims Divisions, & Criminal Divisions

- municipal institutions in the province;

- shop, saloon, tavern, auctioneer, and other licences relating to the raising of revenue for provincial, local, or municipal purposes;

- local works and undertakings other than those relating to steamships or other ships, railways, canals, telegraphs, and other works connecting the province with any other provinces or extending beyond the limits of the province, or with any foreign country;

- incorporation of companies with provincial objects;

- solemnization of marriage in the province;

- property and civil rights in the province;

- administration of justice in the province, including the constitution, maintenance, and organization of provincial courts, both civil and criminal, and including matters of procedure in civil matters in those courts;

- imposition of punishment by fine, penalty, or imprisonment for enforcing any law of the province made in relation to any matter pertaining to those listed in this section; and

- all matters of a merely local or private nature in the province.

For the purposes of this section, it is relevant to discuss the provincial responsibility for public safety issues. There is a wealth of information that relates to the workings of provincial governments in this area, particularly with respect to their ongoing interactions among themselves, but also with respect to their interactions with the federal government in Ottawa. We will content ourselves with a brief overview of those ministries within each of the provincial and territorial governments that have some oversight for law enforcement, policing, or public safety. More detail will be provided on the provincial government level in Chapter 5 ("Policing in a Political Context") and Chapter 8 ("Canadian Government Bureaucracy").

Newfoundland and Labrador   The Justice Ministry is a dual department within the provincial structure. The Minister of Justice is responsible for the administration of court services within the province. This position is also responsible for police protection and any provincial correctional systems.

Within the Justice Ministry, the Attorney General is responsible for administering criminal justice in the province, as well as the prosecution of all offences under the *Criminal Code of Canada* and the statutes of Newfoundland. Also, this minister provides legal services to all government departments and agencies, and represents the Crown where the provincial government's public rights are in question. Finally, the Attorney General offers advice to the government on matters relating to provincial legislation, conducts studies with regard to provincial law reform, and provides legislative drafting services to the government.

Boards, commissions, and agencies in the Justice Ministry include the following:

- Royal Newfoundland Constabulary Complaints Commission

- Legal Aid Commission

- Board of Commissioners of Public Utilities

- Labrador Legal Services

- Human Rights Commission

Prince Edward Island   The Office of the Attorney General includes the Justice and Corrections Division that has responsibility for matters pertaining to policing and law enforcement in this province. Policing within Prince Edward Island is the responsibility of the RCMP, excluding those municipal jurisdictions that maintain their own local police service.

The following boards, agencies, and commissions, all of which have some impact on public safety and security, report to the Attorney General:

- Atlantic Police Academy Advisory Council

- Criminal Code Review Board

- Human Rights Commission

- Victim Services Advisory Committee

New Brunswick   The Department of Public Safety holds responsibility for policing within New Brunswick. The minister ensures that policing standards are established and adhered to within the various municipal police services and works toward improvements in crime prevention activities across the province.

Nova Scotia   The Department of Justice has responsibility in Nova Scotia for promoting the preservation of the peace and providing ongoing efforts at crime prevention. Also, the department works to encourage police efficiency and effectiveness and promotes continuous improvement in the quality of police relationships with various communities within the province. The department currently is divided into the following areas:

- Correctional Services

- Legal Services

- Policy, Planning and Research

- Policing and Victim Services

- Court Services

Of particular interest to those concerned with policing and public safety issues is the Policing and Victim Services division. The policing component of this division has a mandate to

- set policing standards and ensure that police services within the province are delivered in an efficient and effective manner;

- provide a regular system of audits for municipal police services to verify that approved police standards are being met;

- manage contracts for the provision of services by the RCMP, First Nations Policing, and the Atlantic Police Academy;

- provide community-based crime prevention initiatives;

- coordinate police training and education for municipal police services;

- issue licences to companies and individuals engaged in the provision of private security and investigation services; and

- issue Firearms Acquisition Certificates and inspect premises approved to sell firearms.

Quebec   The Minister of Public Security is the provincial Cabinet member responsible to Quebeckers for ensuring a safe and secure society. As such, the Ministry of Public Security oversees policing as well as crime and prevention programs within the province. The minister is directly responsible for the Sûreté du Québec, the provincial police service and, indirectly, other police forces within the province. This ministry also coordinates public complaints about police and work to establish policies and guidelines pertaining to the inspection of police services across the province.

Ontario   In Ontario, matters pertaining to police services and law enforcement fall within the mandate of the Ministry of Public Safety and Security. This ministry replaces the earlier Ministry of the Solicitor General and Correctional Services. The ministry has three main divisions:

- Public Safety

- Policing Services

- Correctional Services

The Policing Services division is responsible for the promotion of policing excellence throughout Ontario. This is achieved through training and education programs, as well as the research and development of policing standards that reflect best practices in all areas of police administration and activity. Furthermore, the ministry provides advisory and liaison services to the police community in Ontario to support adequate and effective police services to all communities. This division includes and audit and investigation capability that is intended to support the goals and objectives of police excellence. Finally, the ministry provides assistance to the policing community through its support of the Criminal Intelligence Service of Ontario.

The Ministry of Public Safety and Security holds responsibility for the following statutes:

- *Ammunition Regulation Act, 1994*
- *Anatomy Act*
- *Christopher's Law (Sex Offender Registry), 2000*
- *Coroners Act*
- *Emergency Plans Act*
- *Fire Protection and Prevention Act, 1997*
- *Lightning Rods Act*
- *Ministry of the Solicitor General Act*
- *Ontario Society for the Prevention of Cruelty to Animals Act*
- *Police Services Act*
- *Private Investigators and Security Guards Act*
- *Public Works Protection Act*

Manitoba   The Department of Justice is responsible for the administration of justice within the province. This department includes the office of the Attorney General. Consistent with its various responsibilities, the department provides funding for several arm's-length boards and commissions that are concerned with public safety and security, including the following:

- Law Reform Commission
- Human Rights Commission
- Criminal Injuries Compensation Board
- Manitoba Police Commission
- Board of Review

While the department continues its role in providing police services within the province, it has recently become more deeply engaged in the formation of community and other partnerships to seek out innovative ways to enhance public safety. This includes expanded efforts at crime prevention, greater concern for the victims of crime, and more attention to alternative approaches to offender management.

Also, the department portfolio includes responsibility for the Law Enforcement Review Agency, created by the *Law Enforcement Review Act*. This agency is designed to provide a service to the citizens of Manitoba with respect to public complaints about the police. The agency has a mandate to register, investigate, and resolve any complaints that come to its attention.

Saskatchewan   The Minister of Justice, who is *ex officio* the province's Attorney General, has a mandate that includes responsibility for the provision of provincial policing services. As a result, the following boards and commissions report to this minister:

- Commission on First Nations and Métis Peoples and Justice Reform
- Human Rights Commission
- Law Reform Commission
- Law Foundation of Saskatchewan
- Saskatchewan Police Commission
- Saskatchewan Police Complaints Investigator

Within the Department of Justice, the Law Enforcement Services Branch works to maintain public order and safety in the province. It accomplishes this mandate by, among other things, ensuring the provision of effective policing programs that observe the rule of law and sustain basic individual rights.

The Law Enforcement Services Branch is specifically responsible for

- administering contracts with the RCMP for the provision of provincial, municipal, and Aboriginal policing services;
- carrying out the day-to-day operations of the Saskatchewan Police Commission which regulates, hears appeals in police discipline cases, and provides training for municipal police services;
- providing centralized training to municipal police services through the Saskatchewan Police College;
- regulating the private security industry; and
- providing a province-wide system of coroners for the investigation of all unnatural deaths and for the conduct of public inquests where the public interest demands an open hearing into the circumstances surrounding a death.

Alberta   The Solicitor General is responsible for ensuring safe communities through the delivery of policing and crime prevention programs within the province. Also, this ministry supports victims of crime during police investigations and criminal court proceedings, and maintains correctional and rehabilitation programs throughout Alberta.

The Law Enforcement Review Board, which provides civilian oversight for policing in Alberta, reports directly to the Solicitor General. In 2002, there was a substantial examination of policing within the province. This review was undertaken by the Policing Alberta MLA Review Committee, made up of members of the Alberta Legislative Assembly. The mandate of this committee was as follows:

To conduct a public consultation on current policing issues and the Alberta Police Act, and to make recommendations for legislative change to the Solicitor General based on the results of the consultation process. *(p. 1)*

The committee undertook the preparation and distribution of a *Policing Review Discussion Paper* that provided a framework for subsequent consideration of the relevant issues and included several key questions for the committee's examination. At the end of its review, the committee produced more than 30 recommendations for the Solicitor General. These recommendations deal with matters relevant to the funding, training, education, and accountability of policing in Alberta. Specific recommendations put forward by the committee include the following:

- require all municipalities to pay for policing;
- divert 25% of provincial fine revenues to a "Policing Support Grant Fund";
- establish a provincial "Policing Support Grant Fund" to provide grants to municipalities;
- establish a provincial "Centre for Excellence" for the provision of police training;
- create an "Alberta Policing Secretariat";
- create a "serious incident review team"; and
- enhance the effectiveness and accountability of police commissions.

British Columbia   Within British Columbia, the Ministry of Public Safety and Solicitor General has overall responsibility for public safety and policing issues. As part of this ministry, the Police Services Division has a mandate to undertake the following:

- ensure the oversight of all policing within the province;
- administer policy relating to policing;
- oversee the organization and financial management of provincial and municipal RCMP policing, providing policing through contracts;
- collect, monitor, and report on provincial crime and police data;
- conduct research, analysis, and interpretation of provincial crime data; and
- ensure that adequate and effective policing is provided across the province.

Yukon   The Department of Justice is responsible for the administration of justice within this territory. Accordingly, it provides all services that may contribute to public safety and security, including appropriate legal services, to the Government of Yukon. The department comprises the following branches:

- Community and Correctional Services
- Community Justice and Public Safety
- Legal and Regulatory Services
- Management Services

The Community Justice and Public Safety Branch provides community justice initiatives, crime prevention programs, and justice-related funding programs and services to the people of Yukon.

Northwest Territories   Within the Northwest Territories the Department of Justice holds the mandate to administer justice, including matters pertaining to policing and corrections. As with other jurisdictions that have a substantial Aboriginal population, the department is particularly aware of its need to observe, and respect, Aboriginal values. As part of the Department of Justice, the Community Justice Division provides assistance to communities in developing alternative approaches to justice issues. This entails seeking appropriate alternatives to the formal justice system that will be consistent with the needs of the community.

Policing for the Northwest Territories is provided by the RCMP. While the RCMP is responsible for controlling its own internal operation, the Minister of Justice works with the command staff of "G Division" every year to establish policing priorities. RCMP officers often work in collaboration with local By-law Community Constables.

Nunavut   The Department of Justice provides legal services to the Cabinet and the Government of Nunavut. It also manages the various programs that are devoted to corrections and community justice within this new territory. Also, this department ensures that the business of the government is conducted in a manner that is consistent with the law and respects the Canadian Constitution and the *Nunavut Act*. The Department of Justice administers the Nunavut Court of Justice and the overall justice system within the territory. Specifically, the department observes the need to respect the role of the Nunavut community in the maintenance of a harmonious and peaceful society, consistent with the principles of Inuit Qaujimajatunqangit. In the area of law enforcement and policing, the department works closely with the Royal Canadian Mounted Police and the other justice-appointed agencies and committees.

Municipal Government Structures

It is always important to remember that municipal governments are the "creatures" of the provinces in which they have come into being. However, there has been a significant growth in the importance of municipalities over the past few decades. This may be clearly seen in the efforts of the Federation of Canadian Municipalities. This organization, which speaks on behalf of municipalities across the country, has evolved into a powerful lobby group seeking to raise the profile and leverage of these corporate entities. This section will not consider municipal jurisdictions in particular detail; however, Chapter 8 ("Canadian Government Bureaucracy") will provide a closer examination of municipal government bureaucracy, particularly with respect to that level's involvement in law enforcement and policing issues, in the cities of Halifax, Montreal, Toronto, Winnipeg, Calgary, and Vancouver.

Ideological Perspectives

More detailed discussion of the specific political parties will be found in Chapter 4 ("The Political Process"). Table 2.4 provides an excellent overview of the various ideological perspectives currently present in the Canadian political landscape.

Table 2.4	The Main Ideologies in Canada		
	Main Value(s)	Social Configuration	Policy Orientation
Conservatism	social order	classes[1]	pragmatic adjustments (e.g., electoral reform)
Liberalism	freedom (and equality)	interest groups (positive)	wage-price control, national energy policy
Neo-conservatism	freedom	interest groups (negative)	deregulation, privatization, monetarism,
Socialism	equality (and freedom)	classes	redistribution, nationalization, industrial democracy,
Populism	equality (independent)	people versus vested interests	social credit: distributive capitalism

1 "The conservative must believe that class distinction can be represented either as a necessary evil or as a social good." Roger Scruton. *The Meaning of Conservatism*. (Totowa, New Jersey: Barnes and Noble Books, 1980), p. 179.

Source: Created by R.C. Nelson.

CONCLUSION

This chapter has provided a broad outline of several essential elements of our Canadian political system. It should be quite clear from this presentation that there is an enormous level of complexity and detail that is embedded in the structures, systems, and services that compose our federalist form of government. The student of Canadian politics has an ongoing challenge to simply stay conversant with the major features of this enterprise.

However, it is important for students, and practitioners, of policing and law enforcement to maintain a fairly high degree of understanding with respect to the larger institutions and agencies within which police organizations operate for the pursuit of public safety. There should be a strong awareness of the impact of these bodies on the functioning, financing, and future direction of policing and law enforcement within their jurisdiction. Policing does not operate within a vacuum, and it is essential that the environment that sustains the realm of policing be comprehensively and continuously scanned for signs of change, conflict, or challenge.

REFERENCES

Alberta Policing Review Committee (2002). Policing review discussion paper.

Baer, Douglas (ed.) (2002). *Political sociology: Canadian perspectives*. Don Mills, Ont. : Oxford University Press.

Bashevkin, Sylvia (1991). *True patriot love: the politics of Canadian nationalism*. Don Mills, Ont. : Oxford University Press.

Bell, David V.J. (1990). "Political culture in Canada." In Whittington, Michael and Glen Williams (eds.). *Canadian politics in the 1990s. 3rd ed.* Scarborough, Ont. : Nelson Canada.

Bell, David V.J. and Lorne Tepperman (1979). *The roots of disunity: a look at Canadian political culture.* Toronto : McClelland and Stewart.

Brooks, Stephen J. (ed.) (1984). *Political thought in Canada: contemporary perspectives.* Toronto : Irwin.

Brym, Robert J. (1989). *From culture to power: the sociology of English Canada.* Toronto : Oxford University Press.

Canada. House of Commons. Special Committee on Reform of the House of Commons (1985). Report of the Special Committee on Reform of the House of Commons (The McGrath Committee). Ottawa : Queen's Printer.

Canada. House of Commons (2001). *A glossary of parliamentary procedure. 3rd ed.* Ottawa : Published under the authority of the Clerk of the House of Commons.

Christian, William and Colin Campbell (1990). *Political parties and ideologies in Canada. 3rd ed.* Toronto : McGraw-Hill Ryerson.

Courtney, J.C. (ed.) (1985). *The Canadian House of Commons: essays in honour of Norman Ward.* Calgary : University of Calgary Press.

Dawson, R. MacGregor (1987). *The government of Canada. 6th ed.* Toronto : University of Toronto Press.

Docherty, David C. (2000). "It's awfully crowded in here: adjusting to the five-party House of Commons." Ottawa : Canadian Study of Parliament Group. Website: **www.study parliament.ca/english/publications_Docherty.htm**

Doern, G. Bruce and Peter Aucoin (eds.) (1979). *Public policy in Canada: organization, process, and management.* Toronto : Macmillan of Canada.

Doerr, Audrey (1981). *The machinery of government in Canada.* Toronto : Methuen.

Dyck, Rand (1993). *Canadian politics: critical approaches.* Toronto : Nelson Canada.

Forsey, Eugene (1974). *Freedom and order: collected essays.* Toronto : McClelland and Stewart Limited.

——— (1984). *How Canadians govern themselves.* Ottawa : Minister of Supply and Services.

Franks, C.E.S. (1987). *The Parliament of Canada.* Toronto : University of Toronto Press.

Grant, George P. (1965). *Lament for a nation.* Toronto : McClelland and Stewart Limited.

Hill, Dilys M. (1974). *Democratic theory and local government.* London : George Allen & Unwin Ltd.

Hockin, Thomas A. (1976). *Government in Canada.* Toronto : McGraw-Hill Ryerson.

——— (ed.) (1977). *Apex of power: the Prime Minister and political leadership in Canada. 2nd ed.* Toronto : Prentice-Hall of Canada Ltd.

Horn, Michiel (1980). *The League for Social Reconstruction: intellectual origins of the democratic left in Canada 1930–1942.* Toronto : University of Toronto Press.

Inside Canada's Parliament: an introduction to how the Canadian Parliament works. Ottawa : Library of Parliament, 2002.

Jackson, R.J. and M.M. Atkinson (1980). *The Canadian legislative system: politicians and policymaking.* Toronto : Macmillan.

Marchak, M. Patricia (1975). *Ideological perspectives on Canada.* Toronto : McGraw-Hill Ryerson.

Meekison, J. Peter (ed.)(1977). *Canadian federalism: myth or reality? 3rd ed.* Toronto : Methuen.

Morton, W.L. (1961). *The Canadian identity.* Toronto : University of Toronto Press.

Pal, Leslie and David Taras (eds.) (1988). *Prime Ministers and premiers: political leadership and public policy in Canada.* Toronto : Prentice-Hall Canada.

Punnett, R.M. (1977). *The Prime Minister in Canadian government.* Toronto : Macmillan.

Quinlan, Don, Terry Lahey, and Mary Jane Pickup (1999). *Government: participating in Canada.* Don Mills, Ont. : Oxford University Press.

Russell, Peter H. (1991). "Can the Canadians be a sovereign people?" *Canadian Journal of Political Science,* Vol. 24, No. 4, pp. 691–709.

Simeon, Richard and Ian Robinson (1990). *State, society, and the development of Canadian federalism*. Toronto : University of Toronto Press.

Smiley, Donald V. (1972). *Canada in question: federalism in the seventies*. Toronto : McGraw-Hill Ryerson.

———— (1987). *The federal condition in Canada*. Toronto : McGraw-Hill Ryerson.

Smith, Jennifer (1999). "Democracy and the Canadian House of Commons at the millennium." *Canadian Public Administration*, Vol. 42, No. 4, pp. 398–421.

Verney, Douglas V. (1986). *Three civilizations, two cultures, one state: Canada's political traditions*. Durham, N.C. : Duke University Press.

Weller, Patrick (1985). *First among equals: Prime Ministers in Westminster systems*. London : Allen & Unwin.

Whitaker, Reginald (1992). *A sovereign idea: essays on Canada as a democratic community*. Montreal : McGill-Queen's University Press.

White, Walter L., Ronald H. Wagenberg, and Ralph C. Nelson (1994). *Introduction to Canadian politics and government. 6th ed.* Toronto : Harcourt Brace Canada.

Wilson, V. Seymour (1993). "The tapestry vision of Canadian multiculturalism." *Canadian Journal of Political Science,* Vol. 26, No. 4, pp. 645–669.

Wittington, Michael S. and Glen Williams (1990). *Canadian politics in the 1990s*. Toronto : Nelson Canada.

WEBLINKS

British Columbia Legislative Assembly

www.legis.gov.bc.ca

Canadian Senate

www.parl.gc.ca

Department of Justice Canada

http://canada.justice.gc.ca/en

Governor General's Office

www.gg.ca

House of Commons

www.parl.gc.ca

Manitoba Legislative Assembly

www.gov.mb.ca/leg-asmb

National Assembly of Quebec

www.assnat.qc.ca

New Brunswick Legislative Assembly

www.gnb.ca/legis

Newfoundland & Labrador House of Assembly

www.gov.nf.ca/hoa

Northwest Territories Legislative Assembly

www.assembly.gov.nt.ca

Nova Scotia Legislature

www.gov.ns.ca/legislature

Nunavut Government

www.gov.nu.ca

Ontario Legislative Assembly

www.ontla.on.ca

Prime Minister's Office

www.pm.gc.ca

Prince Edward Island Legislative Assembly

www.assembly.pe.ca

Privy Council Office

www.pco-bcp.gc.ca

Public Service Commission of Canada

www.psc-cfp.gc.ca

Saskatchewan Legislative Assembly

www.legassembly.sk.ca

Solicitor General of Canada

www.sgc.gc.ca

Supreme Court of Canada

www.scc-csc.gc.ca

Yukon Legislative Assembly

www.gov.yk.ca/leg-assembly

QUESTIONS FOR CONSIDERATION
AND DISCUSSION

1. Does Canada really support the goals of democracy?

    _____

    _____

    _____

    _____

    _____

    _____

2. Does Canada have a class system?

    _____

    _____

    _____

    _____

    _____

    _____

3. What are the key functions of the Prime Minister of Canada?

    _____

    _____

    _____

    _____

    _____

    _____

    _____

4. Is the Senate a valuable government body for today's democracy?

_____

_____

_____

_____

_____

_____

_____

_____

5. What are the benefits of a stable government bureaucracy?

_____

_____

_____

_____

_____

_____

_____

_____

6. How does the mandate of the federal Solicitor General differ from that of the federal Department of Justice?

_____

_____

_____

_____

_____

_____

_____

7. Could Canada have a single, national police service that could provide policing to all provinces, territories, regions, and municipalities? What are the obstacles to such a national police service? What might be some of the benefits of such an approach to policing in Canada?

_____

_____

_____

_____

_____

_____

_____

_____

_____

RELATED ACTIVITIES

1. Review the *Department of the Solicitor General Reference Book* (found on their website, **www.sgc.gc.ca**) and make some determinations about the scope of the department's work and efforts.

2. Select a key issue relevant to the work of the federal Solicitor General or the Department of Justice Canada and follow the evolution of that issue through news releases, statements by the minister, publications, and other elements that might pertain to the matter under review.

3. Review the major websites now available that consolidate all government services relating to public safety and security.

4. Select a provincial, or territorial, jurisdiction and examine the department responsible for public safety and policing in detail. Search that department's website for current information on new programs or initiatives that may be relevant to your studies.

5. Visit Parliament or your provincial legislature and listen to debates among members, particularly those that deal with public safety, security, or policing issues.

6. Study the report of the MLA Policing Review Committee in Alberta and discuss the recommendations that were put forward by this committee.

Canadian Constitutional Issues

It is not in the nature of law to be absolute and unchangeable; it is modified and transformed, like every human work. Every society has its laws, which are formulated and developed with it, which change with it, and which, in fine, always follow the movements of its institutions, its manners, and its religious beliefs.

—Numa Denis Fustel de Coulanges, *The ancient city:*
a study on the religion, laws, and institutions of Greece and Rome

Portrait of Inspector-General John Roche McCowen, Newfoundland Constabulary, circa 1895.

Royal Newfoundland Constabulary.

Learning Objectives

1. Identify the key features of the Canadian Constitution.
2. Outline the process for constitutional change in Canada.
3. Identify the origins of the *Canadian Charter of Rights and Freedoms*.
4. Outline the significance of the *Charter* for policing in Canada.
5. List the seven sections of the *Charter* that are most relevant to policing in Canada.

INTRODUCTION

This chapter will look at the key features of the Canadian constitution. It is essential that readers have a good appreciation of the pertinence of constitutional law to issues of public safety and policing. We will also briefly consider the process of constitutional change in the Canadian context. The *Canadian Charter of Rights and Freedoms* (the *Charter*) will be reviewed as a key component of our nation's commitment to civil liberties, and we will consider the importance of this essential constitutional document to policing in Canada. The *Charter* stands as a major accomplishment in the evolution of Canada; however, it has also had, and continues to have, a profound impact on the way in which policing is provided in this country. Exactly how the *Charter* operates to influence the day-to-day workings of the police will be addressed in this chapter. The seven major sections of the *Charter* that bear directly on policing will be reviewed in some detail.

KEY FEATURES OF THE CANADIAN CONSTITUTION

Several important documents predate Canada's appearance as a "dominion" but still form part of our constitutional heritage. For example, the *Magna Carta,* the *Habeas Corpus Act,* and the *Bill of Rights* of 1689 are all considered components of British constitutional law that have been adopted into Canadian practice. The *Constitution Act, 1867,* addresses issues that are of vital concern to the nature of our political union. Among the most important matters to be dealt with are the following:

- provisions for the union of the colonies;
- outline of the federal government's executive power;
- legislative powers of the federal government;
- provincial constitutions;
- distribution of federal and provincial legislative powers; and
- judicial powers.

The constitutional tradition in Canada fully reflects its British origins. However, it has evolved substantially during the last century and has developed its own unique features. It has also seen the growth of a considerable body of constitutional decision-making issuing from the Supreme Court of Canada.

It is worthwhile for students to reflect on the constitutional relevance of bringing together two distinctive cultures in the creation of Canada. The French reality with its strong Catholic roots, linguistic characteristics, and reliance on a civil law tradition must be understood in contrast to the English presence, characterized by Protestant faiths, a dominant language, and adherence to the Anglo-Saxon common law. The Fathers of Confederation worked to forge these two peoples into a cohesive political entity through appropriate constitutional arrangements. More recent constitutional concerns include the pressing need to address Aboriginal self-government and the protection of fundamental human rights and civil liberties. It is also important to underline the judicial authority that resides in the constitution. The members of the judiciary play an essential role in adjudicating any disputes between the federal government and its provincial counterparts, or among various provincial jurisdictions. The Supreme Court of Canada is central to our constitutional direction (Manning, 1983; Monahan, 1987; Russell, 1982).

Our constitutional arrangements have always been intended to sustain a strong federalist form of government. The need for such a centralizing force was clearly understood by our Fathers of Confederation (though by some more than others). However, the Constitution has permitted, even facilitated, a high degree of flexibility as provincial governments have gained in competence and capacity to assume a greater measure of responsibility.

THE PROCESS OF CONSTITUTIONAL CHANGE IN CANADA

Until 1949, any amendments to the *Constitution Act, 1867,* had to be passed by the British Parliament. Beginning in the 1970s, the issue of constitutional reform became a central topic of debate among Canada's political leaders. The defeat of the sovereignty association initiative in Quebec only served to deepen intense political interest in a renewal of Canadian federalism. The Trudeau government took these matters very seriously and facilitated a meeting of first ministers to discuss some key issues, including the patriation of the constitution, the development of a clear amending formula, and the introduction of a charter of rights for Canadians.

The amending formula that resulted from these meetings requires the agreement of the Parliament of Canada as well as that of at least two-thirds of the provincial legislatures, representing 50 percent of the population. Provinces may choose to "opt out" of specific provisions that they are not in agreement with; however, this would be an extraordinary step, requiring substantial justification.

Efforts have been made to further adjust the constitutional arrangements in Canada in light of outstanding concerns over Aboriginal self-government and the "distinct society" demands of Quebec. The Meech Lake Accord reached in 1987 was not ratified in 1990. The Charlottetown Accord, which was arrived at subsequently by politicians across Canada, was rejected by Canadians in a national referendum held in October 1992. A useful outline of the details of the formulae for constitutional amendment may be found in an essay by the late Senator Eugene Forsey (1996). However, there appears to be a certain weariness with constitutional matters in Canada. This is particularly the case in light of recent public security crises and their troubling economic fallout.

CIVIL LIBERTIES AND THE CANADIAN CHARTER OF RIGHTS AND FREEDOMS

April 17, 1982, is an important date for constitutional law in Canada. This is the date when the *Canada Act* came into force. Schedule B of that legislation is the *Constitution Act, 1982,* and Part I of that act is known as the *Canadian Charter of Rights and Freedoms*. The *Charter* is of considerable importance to law enforcement and policing in Canada and should, therefore, be clearly understood. In combination with the *Canadian Bill of Rights* (1960), which remains in force, the *Charter* represents an essential constitutional document. The *Charter* is part of the "supreme law of Canada" by virtue of its inclusion in the Canadian constitution.

Also, because it is part of the Constitution, any law that is not consistent with the provisions of the *Charter* is, "to the extent of the inconsistency, of no force or effect." This simply means that laws that are found to be contrary to the *Charter* cannot be applied. In effect, a law that offends the *Charter* loses its legal power. For example, a law that permits police to conduct warrantless searches, if found to be contrary to the *Charter,* may be struck down by the courts and rendered ineffective.

The *Charter* is, then, a special piece of legislation. It is a pre-eminent constitutional instrument whose interpretation follows a different path from ordinary statutes (McDonald, 1982).

This legislation provides the fundamental tools for judges and the courts to assess the overall policies and procedures of police departments, as well as the individual actions of police officers as they function in their daily routines. The *Charter* is the litmus test by which the courts measure the appropriateness and acceptability of police approaches to law enforcement. The *Charter* allows the courts to determine if the police are administering justice in a manner that is acceptable within the existing Canadian constitutional framework.

In examinations of the origins of the *Charter,* it has been noted that the late Pierre Elliott Trudeau played a pivotal role in its introduction to Canada (Manning, 1983):

> The Canadian Charter of Rights and Freedoms was really the brain-child of the then Minister of Justice and Attorney General of Canada Pierre Elliott Trudeau. In 1955, in a proposal to the Quebec Royal Commission of Inquiry on Constitutional Problems he stated that Quebec should declare itself ready to accept the incorporation of human rights into the Constitution of Canada and a precise plan should be mapped out for patriating the Canadian Constitution. *(p. 1)*

The introduction of the *Charter* provides a solid basis for making determinations about individual rights in Canada. It fits into an extensive international context that developed during the 1970s and 1980s, when the topic of human rights was significant on the public agenda. Manning (1983) notes the following:

> [O]ur Charter is substantially similar in many ways not only to the American Bills of Rights, but to the United Nations Declaration of Human Rights adopted in 1948, the European Convention on Human Rights and Fundamental Freedoms which came into force in 1953, and the International Covenant on Civil and Political Rights of 1966, to which Canada became a signatory in 1976. Those provisions have been examined by judges of all levels of courts in the United States, by the European Commission of Human Rights and by the European Court of Human Rights and have been applied to fact situations which will arise in Canada. Their decisions will be referred to in the hope that their experience will assist the Canadian judiciary when they come to deal with the application of the Charter. *(p. 3)*

The Canadian judiciary carries a significant responsibility for determining the extent to which the *Charter* will afford protection for individuals (Manning, 1983, p. 2). The *Charter* serves as a symbol of our commitment to human rights, and as a legal mechanism for actually demonstrating that commitment.

The introduction of the *Charter* has certainly increased the prominence of the judiciary, particularly within the Supreme Court of Canada. It has served to focus considerable attention on the 12 justices who constitute the country's highest court (Manning, 1983):

For the first time the explicit guarantees of civil liberties in a constitutional document call upon the judiciary to be the **ultimate expounder** of the meaning and the range of our liberties. It is to the judiciary that the public must now look to determine the scope and extent of their fundamental rights and freedoms. *(p. 21; emphasis added)*

As we can see, it ultimately rests with the Supreme Court of Canada to decide if police powers of detention, arrest, and search and seizure are consistent with the principles of justice within a Canadian context.

The *Charter* plays a leading role in the context of Canadian constitutional law. Since its passage in 1982, the *Charter* has become a critical element in the life and work of the country's highest court, the Supreme Court of Canada. Constitutional law may be seen as the collection and interpretation of laws that guide the relationship between individuals and the state, and between various parts of the state. As one author (Milne, 1982) observes:

The constitution in a federal state forms a master power grid from which all the governmental players get their authority. *(p. 21)*

Because the Constitution is the supreme law in Canada, it is used to determine the validity of all other laws and statutes, federal and provincial.

The *Charter,* incorporating important issues of rights and freedoms, may be employed to resolve central constitutional research problems (Castel and Latchman, 1996). Therefore, anyone interested in learning about police powers and their relationship with individual rights and freedoms will need to know how to study the *Charter* and its application on an ongoing basis. There are several tools available that will assist in this regard (such as Castel and Latchman, 1996, pp. 78–89), and students will certainly benefit from a fairly high degree of familiarity with these aids to legal research.

CONSTITUTIONAL LAW AFFECTING POLICING IN CANADA

Because police officers in Canada are agents of the state with a considerable degree of discretion in the execution of their duties, they have an enormous impact on the liberty of individual citizens. No other public organization can so directly affect the rights and freedoms of individual citizens as the police. Consistent with that impact, police officers have a special legal responsibility that allows them to severely limit our freedoms, and the law extends to them the authority to use whatever means necessary, including lethal force, to constrain our liberties under certain circumstances.

The *Charter* has had a profound impact on police powers in Canada and continues to influence the way police officers and police organizations function across the country. It may be suggested that no single piece of legislation, with the possible exception of the *Criminal Code of Canada,* has greater scope for directing individual officers in their day-to-day functions than the *Charter.*

While the *Charter,* in its entirety, is vitally important to a complete understanding of Canadian constitutional law, seven sections of that document are particularly relevant to police officers in comprehending the nature and scope of their powers. These seven sections warrant special attention in any learning program designed to offer a thorough understanding of arrest, search and seizure, and other police activities aimed at the enforcement of Canadian criminal law. They have each received extensive treatment by the courts, legal scholars, and others concerned with understanding the impact of the *Charter* on policing. Each of these sections will be outlined below; however, readers are encouraged to consult the extensive amount of legal literature available that examines each section in detail. It is useful to become familiar with the wording and focus of each of the highlighted sections in order to facilitate a more extensive examination. It is also important to ensure that any consideration of these *Charter* sections is current and reflects the latest case rulings or statute law. The *Charter* is a special piece of legislation, so it is important to fully understand its elements, including the limitations to which it is subject.

Section 1 of the *Charter* states:

> The Canadian Charter of Rights and Freedoms guarantees the rights and freedoms set out in it subject only to such reasonable limits prescribed by law as can be demonstrably justified in a free and democratic society.

This raises specific questions about the nature of a "free and democratic society" which are beyond the immediate scope of this textbook. However, it is clear that any allowable limits must be reasonable and there must be some demonstration of the justification for those limits. The test of any limits on rights and freedoms will be derived from a wide range of sources, including public opinion, community standards, jurisprudence, sociology, political theory, ethics, and many other sources of insight relevant to matters pertaining to a so-called free and democratic society. Section 1 of the *Charter* deals with the concept of the reasonable limits that may be placed on the protections afforded rights and freedoms in Canada. This section outlines the parameters that can be applied within "a free and democratic society" and, of course, leads to the question of what kinds of reasonable limits might be imposed on such a society.

Section 7 of the *Charter* states:

> Everyone has the right to life, liberty and security of the person and the right not to be deprived thereof except in accordance with the principles of fundamental justice.

This section sets out a relatively broad conceptual framework that has fostered considerable debate over the precise meaning of the terms used to express its meaning. The key individual terms "life," "liberty," and "security of the person" are certainly subject to interpretation, especially when one considers the impact of modern technology and medical ethics, or the influence of global issues like environmentalism, international terrorism, and other matters that could potentially influence our personal security. Also, a precise and unequivocal explanation of "the principles of fundamental justice" is not going to be as readily available as a demonstrable proof of the Pythagorean theorem. However, the fact that these concepts are difficult to articulate in no way diminishes their importance for police officers who wish to learn about the *Charter*.

The context in which Canadian police officers function is defined by the *Charter* and each of the sections holds clues to the interpretations the courts will provide on their specific, day-to-day activities. Also, the justices themselves go to great lengths to provide an articulation of the concepts contained in the *Charter* by using the most compelling research and supporting evidence to clarify their judgements on issues before the court.

Section 7 confirms that the *Charter* views individual rights seriously. It indicates that those rights may only be infringed in ways that can be clearly justified in a society that places high value on the concept of justice. Section 7 states that the protection of life, liberty, and security of the person are things that should matter deeply to us, as citizens and human beings, and that it is only under exceptional circumstances that these rights may be in any way limited. It further tells us that any limitation of these rights must be seen as a "deprivation" and will only be seen as acceptable if the limitation is consistent with notions of fundamental justice that may be said to transcend the practicalities of everyday life.

Section 8 of the *Charter* states:

> Everyone has the right to be secure against unreasonable search or seizure.

This section bears some comparison to the Fourth Amendment of the United States *Bill of Rights,* which reads as follows:

> The right of the people to be secure in their persons, houses, papers and effects, against unreasonable searches and seizures, shall not be violated, and no Warrants shall issue, but upon probable cause, supported by Oath or affirmation, and particularly describing the place to be searched and the persons or things to be seized.

There has been a tradition of turning to American jurisprudence for assistance in making determinations about the application of the *Charter.* The concept of "persuasive authority" should be understood in this context. As Dambrot (1982) explains:

> While it might be desirable to avoid excessive reliance on American cases, it seems inevitable that they will assume an important role in the development of our law, both because of the extensive experience in the United States with an exclusionary rule (while we have virtually none), and because the language of s. 8 seems to have been derived from the American Constitution. *(p. 97)*

On a specific point, when considering an instance of search and seizure that has been approved by an appropriate judicial authority, there is little expectation that it would be found to be unreasonable. Be that as it may, such judicial approval does not exempt a warrant from review (McDonald, 1989):

> A search carried out pursuant to a search warrant obtained without the information in support of the application being in sufficient detail is unreasonable. While in many such cases even without the Charter the search warrant would be held to be invalid, now s.8 of the Charter dictates that the requirements of s.443 of the Code be interpreted in light of s.8. These standards must be considered the minimal standards to be met before a warrant will be properly issued. The right to intrusion must be determined in advance by an impartial judicial officer, and not left to the police. *(p. 251)*

Section 8 has therefore expanded the interest of the Bar in pursuing the "manner" in which a warrant has been obtained. Every aspect or element of the process may be scrutinized to ensure that the requirements of the *Charter,* and by extension, the rights and freedoms of anyone subject to the search and seizure, have been observed.

It is useful to understand the scope of section 8, particularly as it pertains to the guarantee against unreasonable search and seizure. This is a matter of critical importance for police officers. To gain this understanding it is helpful to compare that section to the United States *Bill of Rights* (McDonald, 1982):

There is one feature of the Canadian Charter which differs from the United States Bill of Rights. The latter is directed against the acts of agents of governments; consequently, an act of a private individual who is in no sense the instrumentality of government is held not to be within the constitutional proscription against unreasonable searches and seizures. Section 8 of the Canadian Charter, on the other hand, broadly guarantees "the right to be secure against unreasonable search or seizure", and neither the section on its face nor the context of the section limits the right to security against unreasonable search or seizure to situations in which the search or seizure is by an agent of government…there may be circumstances when evidence has been seized by a private person, the evidence is tendered in evidence for the Crown in a prosecution or for a party in a civil case, and the adversary will object on constitutional grounds to the admission of the evidence. *(pp. 44–45)*

The appearance of section 8 of the *Charter* was really something quite new in Canadian constitutional law. It was, therefore, anticipated as a method for more systematically controlling the application of police powers. The importance of such a constitutional innovation should be understood and emphasized (Manning, 1983):

Section 8 of the Charter of Rights creates a right which was, in the terms used, hitherto unknown and which, when coupled with the broad discretionary remedy of section 24(1) and the exclusionary rule in section 24(2), will have a substantial impact on our jurisprudence. No similar provision existed in the [Canadian] Bill of Rights and while our law has always recognized a method of reviewing actions by police officers in searching and seizing, the control of these acts when found to be unlawful or illegal or even unreasonable has been minimal to say the least. This section was included to be used as a means of controlling not only the issuance of search warrants…but the use of those warrants. The section will also be used to monitor all searches and seizures in relation to regulatory offences both federal and provincial in areas such as customs, excise, wild life protection, environmental protection and health. *(p. 274)*

The scope of section 8 was intentionally designed to be significantly broader than previous constitutional provisions. In looking for a source of this section, it has been suggested that there was ample evidence provided by the McDonald Commission into certain activities of the Royal Canadian Mounted Police (RCMP). The relevance of contemporary Canadian events on the construction of the *Charter* is not to be overlooked, or diminished (Manning, 1983):

Section 8 did not come into being as a protection against the type of abuses that gave rise to the American Fourth Amendment. We must look to our own history in order to gain some insight into the passage of that section. There is no doubt that the framers of section 8 of the Charter had in mind the inquiry into R.C.M.P. activities in the MacDonald [sic] Commission hearings. In the Royal Commission of Inquiry concerning certain activities of the Royal Canadian Mounted Police… there was revealed activities by the police acting without a warrant or acting with a warrant but without any jurisdiction in the justice issuing a warrant. These, undoubtedly, were the kinds of activities that were sought to be curtailed by section 8 and it would be a mistake to ignore the facts revealed in that Report or the public condemnation of those activities. *(pp. 278–279)*

Generally speaking, public inquiries, government law reform studies, and Royal Commissions often have a lasting and meaningful impact on constitutional law, as well as on the judiciary when they are tasked with reviewing *Charter* cases. The importance of section 8 of the *Charter* was certainly predicted early in the day. Its impact on police powers has been considerable as Manning assured us it would be (1983):

> Section 8 promises to be one of the more regularly litigated provisions in the Charter and future decisions should reveal how ready the Canadian judiciary is to follow American authority at a time when there appears to be a retreat from the "blanket exclusionary rule" in the United States. *(p. 322)*

Section 9 of the *Charter* states:

> Everyone has the right not to be arbitrarily detained or imprisoned.

This section has been dealt with in considerable detail elsewhere (McKenna, 2002) and is closely tied to section 10, below.

Section 10 of the *Charter* states:

> Everyone has the right on arrest or detention
>   (a) to be informed promptly of the reasons therefor;
>   (b) to retain and instruct counsel without delay and to be informed of that right; and
>   (c) to have the validity of the detention determined by way of habeas corpus and to be released if the detention is not lawful.

This section is best understood in light of a growing international recognition of human rights and a recognition of the need to formally establish protections against unreasonable arrest and detention. For example, McDonald (1982) is helpful when he makes a comparison between the language of this section and the European Convention on Human Rights:

> The clause in the European Convention alerts us to the possibility that in Canada the word "informed" should be interpreted in a manner which is subjective to the person arrested — in other words, that the person arrested must be "informed" of the reasons for his arrest or detention in a language he understands. Such an argument, if successful, would, from the practical point of view, impose a duty on the police to have interpreters available on a reasonably steady basis, particularly in localities where there are significant numbers of people, such as native people and immigrants, who do not understand the English (or French) language. *(p. 77)*

There is a special onus on police officers to ensure that accused persons have access to legal counsel (McDonald, 1989):

> The provision in the Canadian Bill of Rights has not received extensive judicial interpretation. The leading case is Hogan v. R., [1975] 2 S.C.R. 574 where the majority of the Supreme Court of Canada held that a policeman's refusal to allow the accused access to his counsel, whom he had already retained and who was to his knowledge at the police station where the accused was, violated his right to instruct counsel under section 2(c)(ii) of the Canadian Bill of Rights; however this did not result in the evidence of the result of the breath test he then took being inadmissible. The majority held that common law prevails, that relevant evidence is admissible, no matter how it is obtained, and that the Canadian Bill of Rights has no impact on that rule. *(p. 346)*

The *Charter* adds the requirement that the police are to inform a person of their right to retain and instruct counsel, without delay. McDonald (1989) points out that Canadian law was guided by the important *Miranda* decision in the United States:

> Thus the Canadian Charter falls in line with the position in *Gideon v. Wainwright* [(1964), 378 U.S. 478], in the United States Supreme Court, which held that the failure of a state court to advise a criminal defendant of his right to be represented by counsel is a violation of the Sixth and Fifth Amendments. In *Escobedo v. Illinois* [(1964), 378 U.S. 478], this was held to apply also to the stages of preliminary examination and, still earlier, to police interrogation of an individual held in custody. *(pp. 346–347)*

Section 11 of the *Charter* outlines the rights of an accused. There are nine elements defined in this section, which states:

> Any person charged with an offence has the right
> (a) to be informed without unreasonable delay of the specific offence;
> (b) to be tried within a reasonable time;
> (c) not to be compelled to be a witness in proceedings against that person in respect of the offence;
> (d) to be presumed innocent until proven guilty according to law in a fair and public hearing by an independent and impartial tribunal;
> (e) not to be denied reasonable bail without just cause;
> (f) except in the case of an offence under military law tried before a military tribunal, to the benefit of trial by jury where the maximum punishment for the offence is imprisonment for five years or a more severe punishment;
> (g) not to be found guilty on account of any act or omission unless, at the time of the act or omission, it constituted an offence under Canadian or international law or was criminal according to the general principles of law recognized by the community of nations;
> (h) if finally acquitted of the offence, not to be tried for it again and, if finally found guilty and punished for the offence, not to be tried or punished for it again; and
> (i) if found guilty of the offence and if punishment for the offence has been varied between the time of commission and the time of sentencing, to the benefit of the lesser punishment.

Section 24 of the *Charter* states:

> 1. Anyone whose rights and freedoms, as guaranteed by this Charter, have been infringed or denied may apply to a court of competent jurisdiction to obtain such remedy as the court considers appropriate and just in the circumstances.
> 2. Where, in proceedings under subsection (1), a court concludes that evidence was obtained in a manner that infringed or denied any rights or freedoms guaranteed by this Charter, the evidence shall be excluded if it is established that, having regard to all the circumstances, the admission of it in the proceedings would bring the administration of justice into disrepute.

If a person's rights under the *Charter* have been violated, there is provision for the exclusion of evidence. Much of the groundwork for what is referred to as the exclusionary rule can be found in the Law Reform Commission of Canada's study *The exclusion of illegally obtained evidence: a study paper* (1974), which includes the following observations:

> Canadian law has followed English law: The illegality of the means used to obtain evidence generally has no bearing upon its admissibility. If, for example a person's home is illegally searched — without a search warrant or reasonable and probable cause for a search — the person may sue the police for damages incurred, complain or demand disciplinary action or the laying of criminal charges. But, the evidence uncovered during this search together with all evidence derived from it is admissible. *(p. 5)*

Of course there is an opposing view that endorses fairly blunt treatment for the criminal and resists the tendency to place too many hurdles in the way of the police when they are executing their duties and responsibilities. For example, see the case of *R. v. Honan* (1912), 26 O.L.R. 484, 20 C.C.C. 10 at 16, 6 D.L.R. 276 (C.A.), where Mr. Justice Meredith says:

> The criminal who wheels the "jimmy" or the bludgeon, or uses any other criminally unlawful means or methods, has no right to insist upon being met by the law only when in kid gloves or satin slippers; it is still quite permissible to "set a thief to catch a thief"…

CONCLUSION

We have considered the key features of the Canadian constitution. We have also reviewed the process for undertaking constitutional change as it has developed in Canada. Finally, we have examined the central importance of the *Charter* as a legal document that will guide police powers in the critical areas of arrest, detention, and search and seizure. The origins of the *Charter* as part of the patriated constitution have been reviewed and we have briefly considered the seven key sections of the *Charter* that have a substantial impact on the manner in which police conduct their day-to-day "business."

By considering the pivotal role of the Canadian judiciary and the case law that continues to grow around the key sections of the *Charter*, readers should be able to refine their understanding and knowledge of these provisions. They should also attempt to learn more about updating the developing law that speaks to these issues of police power and its constitutional limits.

Readers are strongly encouraged to delve more deeply into the pertinent sections of the *Charter*. It is essential to have a firm comprehension of this constitutional mechanism for gauging and guiding police powers in Canada.

REFERENCES

Beatty, David (1995). *Constitutional law in theory and practice.* Toronto : University of Toronto Press.

Borovoy, A. Alan (2001). *The fundamentals of our fundamental freedoms: a primer on civil liberties and democracy.* Toronto : Canadian Civil Liberties Education Trust.

Cairns, Alan (1995). *Reconfigurations: Canadian citizenship and constitutional change.* Toronto : McClelland and Stewart.

Castel, Jacqueline R. and Omeela K. Latchman (1996). *The practical guide to Canadian legal research. 2nd ed.* Toronto : Carswell.

Commission of Inquiry Concerning Certain Activities of the Royal Canadian Mounted Police (1981). *Report of the Commission of Inquiry Concerning Certain Activities of the Royal Canadian Mounted Police.* Ottawa : Minister of Supply and Services Canada.

Dambrot, Michael R. (1982). "Section 8 of the Canadian Charter of Rights and Freedoms." In 26 C.R. (3d) 97.

Forsey, Eugene A. (1996). *How Canadians govern themselves. 3rd ed.* Ottawa : Minister of Supply & Services Canada.

Hogg, Peter A. (1997). *Constitutional law of Canada. 4th ed.* [looseleaf]. Toronto : Carswell.

Law Reform Commission of Canada (1974). *The exclusion of illegally obtained evidence: a study paper.* Ottawa : The Commission.

Manning, Morris (1983). *Rights, freedoms and the courts: a practical analysis of the Constitution Act, 1982.* Toronto : Emond-Montgomery.

McDonald, David C. (1982). *Legal rights in the Canadian Charter of Rights and Freedoms: a manual of issues and sources.* Toronto : Carswell.

———— (1989). *Legal rights in the Canadian Charter of Rights and Freedoms. 2nd ed.* Toronto : Carswell.

McKenna, Paul F. (2002). *Police powers I.* Toronto : Pearson Education Canada.

McRoberts, Kenneth and J. Patrick Monahan (eds.) (1993). *The Charlottetown Accord, the referendum and the future of Canada.* Toronto : University of Toronto Press.

Milne, David (1982). *The new Canadian constitution.* Toronto : James Lorimer & Company.

Monahan, Patrick (1987). *Politics and the constitution: the charter, federalism, and the Supreme Court of Canada.* Toronto : Carswell.

———— (2002). *Constitutional law. 2nd ed.* Toronto : Irwin Law.

Monahan, Patrick and Marie Alison Finkelstein (eds.) (1993). *The impact of the Charter on the public policy process.* North York, Ont. : The York University Centre for Public Law and Public Policy.

Reesor, Bayard (1993). *The Canadian constitution in historical perspective.* Scarborough, Ont. : Prentice-Hall Canada.

Romanow, Roy, John Whyte, and Howard Leeson (1984). *Canada...notwithstanding: the making of the Constitution, 1976–1982.* Toronto : Carswell/Methuen.

Russell, Peter H. (1982). *The court and the constitution: comments on the Supreme Court reference on constitutional amendment.* Kingston, Ont. : Institute of Intergovernmental Relations, Queen's University.

———— (ed.) (1987). *Leading constitutional decisions. 4th ed.* Ottawa : Carleton University Press.

———— (1993). *Constitutional odyssey: can Canadians become a sovereign people?* Toronto : University of Toronto Press.

———— (1997). "The end of mega constitutional politics in Canada?" In Guy, James John (ed.). *Expanding our political horizons: readings in Canadian politics and government.* Toronto : Harcourt Brace Canada.

White, Walter L., Ronald H. Wagenberg, and Ralph C. Nelson (1994). *Introduction to Canadian politics and government. 6th ed.* Toronto : Harcourt Brace & Company, Canada.

WEBLINKS

Supreme Court of Canada

www.scc-csc.gc.ca

QUESTIONS FOR CONSIDERATION AND DISCUSSION

1. Does the *Canadian Charter of Rights and Freedoms* enhance or hinder the ability of the police to function?

   _____

   _____

   _____

   _____

   _____

   _____

   _____

2. Is it appropriate for the Supreme Court of Canada to take such a critical role in the determination of *Charter* issues?

   _____

   _____

   _____

   _____

   _____

   _____

   _____

3. How can Canadian police organizations ensure that they are fully aware of the impact of *Charter* decisions?

   _____

   _____

   _____

   _____

_____

_____

_____

_____

_____

4. What, if any, are the differences in the protections provided for individual rights and freedoms between Canada and the United States?

_____

_____

_____

_____

_____

_____

_____

_____

_____

5. Has the introduction of the *Charter* improved individual rights and freedoms in Canada?

_____

_____

_____

_____

_____

_____

_____

_____

_____

_____

RELATED ACTIVITIES

1. Read *Charter* cases in various series of Canadian law reports, e.g., *Canadian Criminal Cases, Criminal Reports,* or other works that focus attention on this area.

2. Peruse your local and/or national newspapers and websites in search of items on the cases relating to the *Canadian Charter of Rights and Freedoms* or items that deal with policing and constitutional rights. Examples of relevant topics could include search and seizure, RIDE programs, the stopping of outlaw motorcycle gang members, and police arrests.

3. Examine the Supreme Court of Canada's website (**www.scc-csc.gc.ca**) and consider individual decisions pertaining to the specific sections of the *Charter* examined in this chapter.

The Political Process

The Lacedaemonian Pedaretus runs for the council of three hundred. He is defeated. He goes home delighted that there were three hundred men worthier than he to be found in Sparta. I take this display to be sincere, and there is reason to believe that it was. This is the citizen.

—Jean-Jacques Rousseau, *Emile or, On Education,* Book I, (translated by Allan Bloom)

Royal Newfoundland Constabulary.

Royal Newfoundland Constabulary patrol officer talking to tourists on Signal Hill, St. John's.

Learning Objectives

1. Identify the 11 political parties currently registered in Canada.
2. Outline the basic elements of Canada's electoral system.
3. Identify the main aspects of party organization in Canadian politics.
4. Outline the basic components of a political campaign in Canada.
5. Describe the general patterns of voting behaviour in Canada.

INTRODUCTION

This chapter will begin by considering the role and function of political parties in Canada. It will look at the key party organizations that have evolved since Confederation and will summarize some of their major contemporary platforms. Next, there will be a review of party organization across Canada. This will involve an analysis of how individual parties operate before, during, and after elections.

The whole complex topic of elections will be reviewed briefly with an emphasis on guiding readers to additional resources that can clarify and elaborate upon this important area of understanding. We will look at the essential elements of an election campaign and will give some consideration to basic patterns of Canadian voting behaviour.

POLITICAL PARTIES IN CANADA

A political party may be seen as any organization that directs its activities toward the goal of occupying the offices that make up the agencies of government. In Canada, there are federal, provincial, and territorial levels of government where political parties may seek formal election. It is rare for political parties to formally run candidates for municipal, or regional, office in Canada. However, municipal and regional politicians frequently have fairly close affiliations with one of the major parties and often graduate to higher levels of elected office following their career in the municipal political arena.

It is interesting to observe that while political parties often coalesce around specific ideological perspectives, a great deal of their members' efforts are aimed at the very pragmatic goal of simply getting elected to political office. Accordingly, the major parties in Canada have organized themselves in a manner that facilitates the straightforward task of gathering votes. They are, in fact, sophisticated machines for securing electoral support. In Ontario, for example, the provincial Progressive Conservative Party has been known as the "Big Blue Machine" in recognition of its formidable powers of organization and efficiency. Professor Paul Thomas (1991) has noted:

> Political parties have long been considered essential to representative and responsible government. Party is the means by which the public puts governments in place and seeks to hold them accountable for their performance. Parties help shape and organize the various options found in society by structuring them in the form of votes and other types of political activity. They give expression to regional and other diversities and integrate them into a definition of national interest. *(p. 191)*

To be recognized as an official political party in Canada, the group must be registered with Elections Canada. As of the beginning of 2003, the following 11 registered political parties existed in Canada:

- Bloc Québécois
- Canadian Action Party
- Canadian Reform Conservative Alliance
- Communist Party of Canada
- Green Party of Canada
- Liberal Party of Canada

- Marijuana Party
- Marxist-Leninist Party of Canada
- Natural Law Party of Canada
- New Democratic Party
- Progressive Conservative Party of Canada

Bloc Québécois

This is a federal political party dedicated to Quebec sovereignty. Its origins can be traced back to the resignation of federal MP Lucien Bouchard, who had been the Minister of the Environment and the Minister responsible for La Francophonie. He left the Conservative caucus of Brian Mulroney and sat as an independent. With the death of the Meech Lake Accord, which sought to clarify Quebec's place in Confederation, a considerable amount of interest was generated for Quebec independence. In this context, Lucien Bouchard and several like-minded colleagues were instrumental in the creation of the Bloc Québécois.

Under Bouchard's leadership the party held a founding convention in June 1991. In September of that year it registered as an official party and ran candidates for the 1993 election. The results of this election were profound. With the Bloc Québécois winning 54 seats, Lucien Bouchard became the leader of the official Opposition in the House of Commons. Consistent with the party's mandate to pursue independence, a referendum on Quebec was held on October 30, 1995. The federalist "No" side secured 50.5 percent of the vote while the "Yes" side won 49.5 percent support. A very narrow margin preserved Canada intact, for the time being. Within the Province of Quebec, the Parti Québécois, a mirror of the federal Bloc, won election as the provincial government in November 1998. "Sovereignty association" or independence for Quebec continues to be the strategic goal of both the Bloc Québécois, at the federal level, and the Parti Québécois, at the provincial level.

Canadian Action Party

The Canadian Action Party, formed in 1997, seeks to re-establish Canada as an independent country with its own political and economic sovereignty. It opposes what it terms the ascendancy of "corporate rule," as well as the colonization and absorption of Canada by the United States. It has taken a stand against corporate globalization and opposes the purposes of the World Trade Organization (WTO). The Canadian Action Party calls for the abrogation of the Free Trade Agreement with the United States as well as the North American Free Trade Agreement (NAFTA).

This party is committed to greater access to information so that Canadians are more aware of public policies. In the party's view, this knowledge will allow the people to participate more fully in the political, economic, social, cultural, and general well-being of the country. It believes that the state has a responsibility to provide citizens with certain essential services, and for that purpose it seeks to reform the Canadian banking system and use the Bank of Canada to provide a greater degree of fiscal responsibility. It supports the closure of the World Bank and the International Monetary Fund, which it sees as institutions dedicated to the re-colonization of developing countries and creating conditions that will leave them prey to large multinational corporations.

In the area of law and order, the Canadian Action Party seeks to reduce fear of crime to the maximum extent possible in our country, especially for women.

Canadian Reform Conservative Alliance

The short name of this party is the Canadian Alliance, and it is an outgrowth of the previous Reform Party of Canada that developed in Western Canada under the leadership of Preston Manning and his successor, Stockwell Day. It presents itself as a broad coalition of citizens who are seeking an alternative to previous governments and aims to ensure greater accountability and responsibility in the political arena.

The origins of the earlier Reform Party have been addressed in detail by other authors (Dobbin, 1991; Flanagan, 1995; Harrison, 1995; and Sharpe and Braid, 1992). However, in 1998, a number of individuals interested in the advancement of a broad-based, democratic conservative movement in Canada met at a national convention to formulate future plans. This meeting led to an event in February 1999 known as the United Alternative Convention that was intended to forge the basis of a new political party in Canada. Following the work of the members at this convention (which included members of the Reform Party) a second convention held in January 2000 led to the creation of the Canadian Reform Conservative Alliance. The Canadian Alliance thus became a new political party with a national base, a detailed policy declaration, and a new constitution.

The Canadian Alliance has outlined several policy declarations in the area of law and order that may be relevant to police organizations across the country. Overall, the party supports full funding for law enforcement agencies and is committed to substantial reform of the Canadian criminal justice system, including increased accountability for young offenders.

Communist Party of Canada

This party, which drew its inspiration from the Russian Revolution of 1917, came into being in Guelph, Ontario, in May 1921. Not to be confused with the Communist Party of Canada (Marxist-Leninist), this is a distinct political entity that has been running candidates in federal and provincial elections for many years.

Green Party of Canada

Also known as the Canadian "Greens," this party is linked with a global environmental movement that is deeply committed to creating a green society. Accordingly, it works to encourage the evolution of social and political structures that will support Green Values, including

- non-violence and military disarmament;
- ecologically sustainable economy;
- preservation and restoration of ecosystem diversity;
- eco-feminism;
- cultural and multiracial diversity;

- consensual decision-making process;

- decentralized decision making; and

- wholistic community through personal and global responsibility.

Liberal Party of Canada

The origins of the Liberal Party of Canada can be found in the earliest days of settlement in Lower Canada (now Quebec) and Upper Canada (now Ontario). Reformers who opposed the dominant ruling class theory struggled to gain recognition for their political goals. In Lower Canada, a group known as the "Reformers" were led by Louis-Joseph Papineau. In Upper Canada, the Reformers were led by William Lyon Mackenzie. Both groups were attempting to erode the special powers and privileges of the ruling oligarchies in their respective jurisdictions: the "Family Compact" in Upper Canada and the "Château Clique" in Lower Canada. With the passage of the *Act of Union* in 1834, these two jurisdictions were combined to form a single legislative entity, the Province of Canada. In attempting to bring responsible government to the people of Canada, the leaders of the Canadian Reformers, Louis-Hippolyte Lafontaine and Robert Baldwin, campaigned and were elected in 1848. This dual French and English leadership became symbolic of the Liberal Party's commitment to the union of Quebec and Ontario.

The Conservatives, under John A. Macdonald, held power at the federal level from the time of Confederation until 1896, with the exception of the years 1873 to 1878. In November of 1873, the Macdonald government was compelled to resign in the face of the Pacific Scandal, and the Liberals, under Alexander Mackenzie, were elected to power with 133 seats. When Wilfrid Laurier became the leader of the Liberals, he was able to bring together members of the Liberal Party of Canada for their first national convention in 1893. In 1896, the Laurier Liberals won 117 seats and were able to form the federal government.

In the area of justice, the Liberal Party of Canada has recently identified several key issues that have received their attention as the governing party. It has acknowledged that, in spite of declining crime rates across Canada, people continue to be concerned about crime and disorder. To address this concern, the Liberals identify the following actions that their government has taken, as well as the directions they plan to pursue:

- creating a National Crime Prevention Strategy, which has funded a wide range of projects in 435 communities;

- launching a comprehensive review of the youth justice system in Canada that will address the prevention of youth crime, work toward the introduction of meaningful consequences for youth crime, and place an emphasis on rehabilitation and reintegration;

- placing greater emphasis on the challenges of organized crime through investments in anti-smuggling initiatives, additional funding for the Royal Canadian Mounted Police (RCMP), and assisting police organizations in the development of approaches for dealing with the new technologies employed by criminals;

- making significant financial investments in information systems that will link police organizations across the country;

- creating a new DNA data bank;

- investing in a new policy centre for victims' issues;

- broadening and sustaining community crime prevention efforts;

- targeting the proceeds of crime;

- implementing a strategy that will reduce the supply and the demand for illegal drugs; and

- strengthening anti-gang legislation.

Thorburn (1996) calls the Liberal Party "the government party" of Canada. It has held federal power during significant periods in Canadian history, and it currently dominates the federal political scene.

Marijuana Party

As its name implies, the Marijuana Party is dedicated to the legalization of marijuana use in Canada. It is, therefore, a special-interest political party with a fairly closely defined focus. Surprisingly, the party is planning to run candidates in all 301 Canadian ridings to support its aspirations to pursue the emancipation of cannabis.

Marxist-Leninist Party of Canada

This is the Communist Party of Canada (Marxist-Leninist). For the purposes of differentiating it from the Communist Party of Canada, it is referred to as the Marxist-Leninist Party of Canada. Guided by Hardial Bains for several decades through 1997, it is dedicated to the political principles of Marx and Lenin, and seeks the overthrow of the bourgeois, capitalist system in Canada. It has a very active website that includes an Internet edition of *The Marxist-Leninist Daily,* which provides detailed statements about its position on major issues confronting Canadians, both at home and on the international scene.

Natural Law Party of Canada

The Natural Law Party has counterparts in other countries, including the United Kingdom and the United States. Essentially, it promotes the advancement of human life through the application of the principles of Transcendental Meditation (TM), a system of stress reduction most closely associated with the work of the Maharishi Mahesh Yogi. Some people would refer to such an organization as a "fringe" political party.

New Democratic Party

The New Democratic Party (NDP) came into being in 1961 during a convention in Ottawa. It combined members of the Co-operative Commonwealth Federation (CCF) and affiliates of the Canadian Labour Congress. The NDP carries on traditions of the democratic socialist movement in Canada, and its policies and platforms have been substantially informed by several key events in Canadian history. The Winnipeg General Strike of 1919 was particularly important in setting the agenda for any political party that sought to support the continuing efforts of the labour movement in Canada for greater equality and recognition.

Some of the key issues that currently drive the NDP include the following:

- *democracy and human rights,* including Aboriginal treaty rights, equality for persons with disabilities, gay and lesbian rights, and women's equality;

- *infrastructure,* including the linkage of Canadians across the country, and a commitment to urban transit in the twenty-first century;

- *culture,* including a strong Canadian Broadcasting Company (CBC), preservation of Canadian cultural identity, and reductions in the ongoing concentration of Canadian media;

- *health care,* fighting the privatization of health care, and fighting for lower prescription costs;

- *employment,* training to meet the needs of the new economy, improved conditions for the unemployed, job security;

- *communities,* including improvements in federalism, addressing the crisis of homelessness, improving immigration, building safer communities, and making a commitment to livable cities;

- *education,* including the problems of high tuition and student debt;

- *economy,* including a bottom-up approach to community development, securing quality jobs, fair taxation, and fair trade;

- *environment,* including national leadership for a "green" approach to the Canadian economy;

- *international issues,* including a stronger defence policy that allows for increased immigration;

- *First Nations,* including respect for Aboriginal treaty rights, and improved health care;

- *children,* including increased investments in early child care and addressing the problems of child neglect; and

- *natural resources,* including support for family farms, East Coast fisheries, and improvements in the quality of drinking water.

Progressive Conservative Party of Canada

The Progressive Conservative Party began in 1854, with the Conservative Party of John A. Macdonald. The party originally represented a coalition of four pre-Confederation groups, namely Tories and Moderates from Upper Canada (now Ontario) as well as several English businessmen and French-speaking conservatives from Lower Canada (now Quebec). This grouping was not entirely stable, but in contrast to other parties vying for power at the time, it coalesced into a relatively homogeneous and organized political party. It was this level of cohesion and organization that propelled the Conservatives, under John A. Macdonald, into the first government of the Dominion of Canada. Furthermore, the Conservative Party held onto federal power from Confederation until the "Pacific Scandal" caused the party to lose control of government and led to the election of the Liberals.

The Progressive Conservative Party of Canada, as it became known in 1942, has been labeled as "the opposition party" (Thorburn, 1996) largely because it has spent so many years as the official Opposition to the ruling Liberal Party. The Conservatives seem to function quite effectively in that role. Finally, efforts at a marriage between the Progressive Conservative Party of Canada and the Canadian Alliance have been, to date, largely unsuccessful. Table 4.1 presents the standing of parties in the House of Commons as of September 18, 2002.

The party system in Canada is seeing a fairly high degree of transformation, and there may be several reasons for this change. One is an increasing level of public cynicism with regard to politicians and their behaviour. There is also an emergence of more sharply defined regional concerns. This has been exemplified most dramatically in Quebec with the rise of the Bloc Québécois upon a platform of separation. The appearance of the Reform Party was also symptomatic of a regional distrust of conventional political parties. Even the NDP has not been immune from a growing doubt among Canadian democratic socialists that it can adequately represent the interests of labour and other social democrats in their current configuration. Much is happening on the political scene and many citizens, rather than joining the mainstream political parties, have begun to find their way into any number of special interest groups or new social movements in an attempt to have their policy, or ideological, aspirations met (Tanguy, 1994).

Many political scientists refer to the "brokerage theory" of parties in Canada. This theory understands political parties to be essentially mediators of issues and ideas within the political arena. Or, as outlined by Rand Dyck (1993):

> The essence of this interpretation is that, given the multiple cleavages in Canadian society and the function of parties to aggregate interests, political parties in Canada should be conciliators, mediators, or brokers among the cleavages already identified, that is, regions, ethnic and linguistic groups, classes, religions, ages, and genders. The theory suggests that maximizing their appeal to all such groups is not only the best way for parties to gain power, but in the fragmented Canadian society this approach is also necessary in order to keep the country together. Thus, in their search for power, parties should act as agents of national integration and attempt to reconcile as many divergent interests as possible. *(p. 272)*

PARTY ORGANIZATION

All political parties are concerned with creating the best structure possible to allow them to pursue their political goals. Dyck (1993) has differentiated between a "mass" style of party organization and a "cadre" approach to party organization. The mass style of party organization is clearly aiming at the broadest kind of representation and is more fully informed by democratic principles. The cadre style is characterized by a more elite understanding of the mechanics of party organization. By accepting this dichotomy of party organizational structure, it is possible to consider each party according to the following elements:

Membership.   A political party will demonstrate its mass or cadre characteristics through its approach to membership. The mass party will seek to have a broad membership and will seek to expand its membership rolls on a continuous basis. The cadre party will focus on a small, elite group to form the core of its membership.

Table 4.1 | House of Commons Party Standings, 37th Parliament

Updated on September 18, 2002

Province	Liberal	Canadian Alliance	Bloc Québécois	NDP	Progressive Conservative[1]	Ind.[2]	Vacant	Total
Alberta	2	23			1			26
British Columbia	6	26		2				34
Manitoba	5	3		4	2			14
Newfoundland and Labrador	4				3			7
New Brunswick	6			1	3			10
Northwest Territories	1							1
Nova Scotia	4			3	4			11
Nunavut	1							1
Ontario	99	2		2				103
Prince Edward Island	4							4
Quebec	35		35		1	2	2	75
Saskatchewan	2	9		2		1		14
Yukon	1							1
Total	170	63	35	14	14	3	2	301

1 The Speaker recognized the 12 members of the Progressive Conservative Party as members of the PC/DR Coalition (Monday, September 24, 2001). End of the PC/DR Coalition (April 10, 2002).

2 The Speaker recognized the 8 independent members who comprise the Democratic Representative Caucus as members the PC/DR Coalition (Monday, September 24, 2001). Mr. Lunn excluded from the PC/DR Coalition as of November 19, 2001. End of the PC/DR Coalition (April 10, 2002).

Source: Information and Documentation Branch, Library of Parliament, Elections Canada, Enquiries Unit.

Leadership.   The mass party will choose its leader through a convention where every member has an opportunity to vote and, once a leader has been selected, will ensure that there is a high level of accountability to the party on the part of that leader. The cadre party will typically allow the parliamentary party or caucus to select the leader and there will be far less formal accountability once that selection has been made.

Finance.   The mass party is normally financed through a membership fee, as well as the voluntary contributions from its extensive membership. The cadre party typically operates on the strength of funding from a small number of large (often corporate) contributors.

Policy-making.   Broad and extensive member input is sought by the mass party in the process of formulating its policy positions. Cadre parties, on the other hand, will rely upon the parliamentary party to develop their policy declarations.

General structure and operations.   A mass party exemplifies a more democratic overall structure and operates with a great deal more responsiveness to its members on a day-to-day basis. It normally has an established constitution that sets out in detail the policies, procedures, and practices of the party. The cadre party is usually run by an elite group and may not strictly adhere to its official constitution.

While each of the recognized political parties in Canada has unique characteristics, they all organize themselves in ways that to a greater or lesser extent reflect each of these elements. By and large, Canada is characterized by mass party organization; however, there are, perhaps inevitably, cadre elements in most of the mainstream parties that have formed the various federal and provincial governments in Canada.

ELECTIONS

Canada has a parliamentary system of government that is, as noted in Chapter 1, derived in large measure from the British form of government. One key feature of such a form of government is the institution of free elections for the selection of candidates for government office. The singular act of casting a ballot is central to the Canadian democratic enterprise. Dyck (1993) makes the following observation:

> By exercising their franchise, voters legitimize the power of those elected by agreeing to be bound by their decisions. This activity presumably also generates popular support for the political system as a whole. Elections similarly help to integrate the system by putting everyone in it through a common national experience. *(p. 232)*

Canada currently has 301 electoral districts. Each district is authorized to elect one person to represent that constituency (or riding) in the House of Commons. Those persons elected are known as members of Parliament (MPs) and they hold office until a subsequent election is called.

Elections Canada is the federal government department that is responsible for the administration of the electoral process in this country. It is an independent, non-partisan agency that reports directly to the Canadian Parliament through the Chief Electoral Officer. It operates to ensure that Canadians are able to exercise their voting rights during elections and referendums through processes that are both open and impartial. The mandate of Elections Canada is to exemplify the highest level of excellence in electoral affairs.

It operates to provide a professional electoral process that meets the needs of the Canadian electorate, as well as our legislators, in an effective, efficient, and economical manner. Furthermore, this agency is responsible for the administration of the *Canada Elections Act.*

The federal election of October 25, 1993, was an extremely significant one within Canadian history. The Progressive Conservatives, who had been the ruling majority party, were reduced to a mere two seats in the House of Commons. This amounted to a staggering repudiation of the party and was the worst electoral defeat in Canada's history. Also, the New Democratic Party plunged from 44 seats in the House of Commons to only eight. The Liberal Party under Jean Chrétien's leadership succeeded in winning 178 seats in Parliament, a clear and resounding majority. Added to this dramatic reversal of fortunes was the strong showing of the Bloc Québécois. This party of Quebec separation won 54 seats and therefore became the official Opposition in the House of Commons. Finally, the Reform Party (the precursor of the Canadian Alliance) won 52 seats, underscoring the high level of dissatisfaction with "politics-as-usual" on the part of the Western provinces.

Although voting takes place between 8:00 a.m. and 7:00 p.m., it should be remembered that Canada encompasses six time zones; therefore, the *Canada Elections Act* requires that the results from other provinces cannot be publicized in any province whose polls are still open.

Appointed elections officers known as returning officers divide each constituency into polling divisions, or polls, that consist of approximately 250 voters each. Following the conclusion of an election, returning officers are responsible for validating the results of the election. If necessary, judicial recounts are conducted. Once all these steps have been completed, returning officers declare the winning candidates.

Secret ballots were introduced in 1874. Therefore, the first two federal elections in the Dominion of Canada were oral and hence were subject to bribery and intimidation. The Inuit were accorded the right to vote in 1950. Indians living on reserves were able to vote after 1960. In 1970 the age of majority was changed from 21 to 18, and British subjects who were not Canadian citizens lost the right to vote in our elections in 1975. Currently, those who are ineligible to vote include

- Chief Electoral Officer

- Assistant Chief Electoral Officer

- returning officers (except in the event of a tie)

- prison inmates

- persons convicted of corrupt or illegal electoral practices

Election Campaigns

Political campaigns are enormously complex and important undertakings. They ultimately guide voters in their choice of elected officials who will control the levers of government. Therefore, political parties spend a great deal of time planning how they will approach the campaign process. The deliberations of each political party must include decisions about the kind of advertising they will employ on behalf of their candidates, as well as the forms of media that will be used to get their messages across to the Canadian voting public. For example, campaign managers have to think about the amount of radio, television, and print exposure the party will seek to convey its particular message and to promote its own candidates. Of course, it is not surprising that the Internet has taken on an enormous role in this process today.

Each party has something akin to a finance committee that makes fundamental determinations about the amount of money available within each constituency. Party headquarters may decide that it will supplement these funds in support of a particularly promising candidate or where the race among different candidates is especially close. In looking at where money will be spent, party headquarters must make some difficult decisions to maximize the party's financial resources. Obviously, media time can be a very expensive proposition, so parties are cautious with their commitment to modes of communication that do not show their candidate to best advantage. Furthermore, campaign managers realize that they must concentrate their efforts on undecided voters, and this means managers must gain the greatest exposure for their candidate with these people. Through extensive and sophisticated polling techniques, parties are able to diagnose the mood of potential voters and plan strategies that will attract them to their party.

Political campaigns require small armies of volunteers and party officials to ensure that their messages saturate the constituencies in which they are running candidates. Campaign workers are active until the day of the election, distributing flyers, conducting polls, and canvassing ridings on behalf of their party.

At the federal level, election campaigns have become increasingly sophisticated, with substantial numbers of people involved in the multi-faceted demands of promoting the party's leader, its policies, and the various local candidates.

Canadian Voting Behaviour

There are many factors that must be taken into account when examining how, and why, a voter makes an electoral decision. Frequently, there are significant national issues that tend to drive or polarize the Canadian electorate. For example, the National Energy Policy (NEP) under the Trudeau government, the free trade issue during the Mulroney government, and universal health care are among the types of issues that may galvanize a particular election. Because political issues may arise with little, or no, predictability, it is not surprising that voter behaviour is also hugely unpredictable. World issues such as the September 11, 2001, attacks, the Israeli-Palestinian crisis, and the Kyoto Accord may drive Canadian politics in ways that neither politicians nor voters can anticipate. If we add to this the impact of the various distinctions, or "cleavages," that were discussed in Chapter 1, it is not surprising that voter behaviour is a most mysterious puzzle.

There are several myths surrounding voter behaviour and its predictability. Rand Dyck (1993) has made the following observations on those myths:

> Research by political scientists into electoral behaviour has shattered many myths. These include the myth of the rational, well-informed voter, the myth of the voter securely attached to one party, the myth of the voter who weighs party positions on various issues before deciding how to cast his or her ballot, the myth that the voter identifies with the ideological position of a particular party, and the myth that most voters make up their mind long before the election is called. *(p. 254)*

Federal elections tend to attract from 70 to 75 percent of the electorate, though very recent elections have attracted even lower levels of participation. This means that many Canadians choose not to vote, which may seem ironic given that one of the hallmarks of a representative democracy is the capacity to exercise one's right to vote. To get some perspective on the number of people who were qualified to vote, the actual number of ballots cast by qualified voters, and the resulting percentage of voter participation, Tables 4.2, 4.3, and 4.4 are particularly useful.

Table 4.2	Federal General Elections, by Electors			

Last modified: January 4, 2002

Electors on the Lists	1988	1993	1997	2000
Canada	17,639,001	19,906,796	19,663,478	21,243,473
Newfoundland and Labrador	384,236	419,635	407,109	405,210
Prince Edward Island	89,546	99,645	97,802	103,034
Nova Scotia	644,353	707,202	677,164	694,984
New Brunswick	508,741	562,128	551,530	571,569
Quebec	4,740,091	5,025,263	5,177,159	5,542,169
Ontario	6,309,375	7,266,097	7,115,785	7,713,744
Manitoba	729,281	791,374	758,526	786,309
Saskatchewan	675,160	704,248	679,806	698,145
Alberta	1,557,669	1,851,822	1,811,413	2,094,001
British Columbia	1,954,040	2,420,709	2,332,083	2,574,322
Yukon	16,396	20,565	19,934	20,901
Northwest Territories	30,113	38,108	35,167	24,716
Nunavut	—	—	—	14,369

— data unavailable, not applicable, or confidential

Source: Elections Canada.

Table 4.3	Federal General Elections, by Ballots Cast			

Last modified: January 4, 2002

Total Ballots Cast	1988	1993	1997	2000
Canada	13,281,191	13,863,135	13,174,698	12,997,185
Newfoundland and Labrador	257,793	231,424	224,722	231,178
Prince Edward Island	75,986	72,973	71,171	74,888
Nova Scotia	481,682	457,610	470,272	437,375
New Brunswick	386,201	391,247	404,921	387,178
Quebec	3,562,777	3,873,050	3,792,870	3,552,551
Ontario	4,706,214	4,918,819	4,664,525	4,474,001
Manitoba	544,756	543,339	479,110	490,083
Saskatchewan	525,219	488,755	444,004	435,079
Alberta	1,167,770	1,206,871	1,059,348	1,259,794
British Columbia	1,538,628	1,640,614	1,529,139	1,621,101
Yukon	12,849	14,471	13,915	13,272
Northwest Territories	21,316	23,962	20,701	12,912
Nunavut	—	—	—	7,773

— data unavailable, not applicable, or confidential

Source: Elections Canada.

Table 4.4	Federal General Elections, by Voter Participation			
Last modified: January 4, 2002				
Voter Participation (%)	1988	1993	1997	2000
Canada	75.3	69.6	67.0	61.2
Newfoundland and Labrador	67.1	55.1	55.2	57.1
Prince Edward Island	84.9	73.2	72.8	72.7
Nova Scotia	74.8	64.7	69.4	62.9
New Brunswick	75.9	69.6	73.4	67.7
Quebec	75.2	77.1	73.3	64.1
Ontario	74.6	67.7	65.6	58.0
Manitoba	74.7	68.7	63.2	62.3
Saskatchewan	77.8	69.4	65.3	62.3
Alberta	75.0	65.2	58.5	60.2
British Columbia	78.7	67.8	65.6	63.0
Yukon	78.4	70.4	69.8	63.5
Northwest Territories	70.8	62.9	58.9	52.2
Nunavut	—	—	—	54.1

— data unavailable, not applicable, or confidential

Source: Elections Canada.

The impact of gender on voting behaviour has also become an important area of study, including the impact of feminism on party politics (Young, 2000; Cross, 2001).

CONCLUSION

This chapter has attempted to provide a basic outline of the party system as it operates in the Canadian context. There has been some attention given to the 11 political parties currently registered by Elections Canada, and a brief accounting of some of their key policy positions.

We have looked at the organizational structure of political parties in Canada and have seen more similarities than differences in their overall approach to doing business. The electoral system in Canada has been reviewed in outline and the key features of this complex matter have been identified. The substantial references in this chapter allow for a more detailed study in this area of Canadian politics.

The elements of an electoral campaign have been considered, along with the challenging matter of Canadian voter behaviour. This chapter addresses some important aspects of the critical machinery of Canadian democracy. It is essential that the key components of this machinery be understood as they continue to evolve in ways that, hopefully, will secure for the Canadian public the best and most responsive representation and leadership in political matters.

REFERENCES

Archer, Keith (1990). *Political choices and electoral consequences.* Montreal : McGill-Queen's University Press.

Bakvis, Herman (ed.) (1992). *Canadian political parties: leaders, candidates and organization.* Toronto : Dundurn Press.

Black, Jerome H. (1984). "Revisiting the effects of canvassing on voter behaviour." *Canadian Journal of Political Science,* (June), pp. 351–374.

Brodie, Janine and Jane Jenson (1988). *Challenge and change: party and class in Canada revisited.* Ottawa : Carleton University Press.

Brook, Tom (1991). *Getting elected in Canada.* Stratford, Ont. : Mercury Press.

Cairns, Alan C. (1968). "The electoral system and the party system in Canada, 1921–1965." *Canadian Journal of Political Science,* Vol. 1, No. 1 (March), pp. 55–80.

Canada. Royal Commission on Electoral Reform and Party Financing (1991). *Reforming electoral democracy.* Toronto : Dundurn Press.

Carrigan, D.O. (1968). *Canadian party platforms: 1867–1968.* Toronto : Copp Clark.

Carty, R.K., Lynda Erickson, and Donald Blake (eds.) (1992). *Leaders and parties in Canadian politics: experiences of the provinces.* Toronto : Harcourt Brace Jovanovich Canada.

Carty, R.K., William Cross, and Lisa Young (2000). *Rebuilding Canadian party politics.* Vancouver : University of British Columbia Press.

Christian, William and Colin Campbell (1995). *Political parties, leaders and ideologies.* Toronto : McGraw-Hill Ryerson.

Clarke, Harold et al. (1996). *Absent mandate: Canadian electoral politics in an era of restructuring. 3rd ed.* Toronto : Gage Publishers.

Courtney, John C. (1978). "Recognition of Canadian political parties in Parliament and the law." *Canadian Journal of Political Science,* Vol. 11, No. 1 (March), pp. 33–66.

———— (1995). *Do conventions matter? Choosing national party leaders in Canada.* Montreal : McGill-Queen's Press.

Cross, William (1996). "Direct election of provincial party leaders, 1985–1995: the end of the leadership convention?" *Canadian Journal of Political Science,* Vol. 24, No. 2, pp. 295–315.

———— (2001). *Political parties, representation, and electoral democracy in Canada.* Toronto : Oxford University Press.

Docherty, David C. (1998). "Parliaments and government accountability." In Westmacott, Martin and Hugh Mellon (eds.). *Public administration and policy: governing in challenging times.* Toronto : Prentice-Hall.

Dyck, Rand (1993). *Canadian politics: critical approaches.* Scarborough, Ont. : Nelson Canada.

Elections Canada (1996). *Strengthening the foundation: Canada's electoral system: annex to the report of the Chief Electoral Officer of Canada on the 35th general election.* Ottawa : Chief Electoral Officer of Canada.

———— (1997). *A history of the vote in Canada.* Ottawa : Canadian Government Publishing.

Engelmann, F.C. and M.A. Schwartz (1975). *Canadian political parties: origin, character, impact.* Scarborough, Ont. : Prentice-Hall Canada.

Everitt, Joanna and Brenda O'Neill (eds.) (2002). *Citizen politics: research and theory in Canadian political behaviour.* Don Mills, Ont. : Oxford University Press.

Heggie, Grace F. (1977). *Canadian political parties, 1867–1968: a historical bibliography.* Toronto : Macmillan of Canada.

Hurtig, Mel (1992). *A new and better Canada: principles and policies of a new Canadian political party.* Toronto : Stoddart Press.

Johnson, R. et al. (1992). *Letting the people decide: the dynamics of a Canadian election.* Stanford, Calif. : Stanford University Press.

Johnston, Paul and Harvey Pasis (1990). *Representation and electoral systems: Canadian perspectives.* Scarborough : Prentice-Hall Canada.

Johnston, Richard (1985). "The reproduction of religious cleavage in Canadian elections." *Canadian Journal of Political Science,* (March), pp. 99–117.

Paltiel, K.Z. (1970). *Political party financing in Canada.* Toronto : McGraw-Hill Ryerson.

Pammett, Jon (1987). "Class voting and class consciousness in Canada." *Canadian Review of Sociology and Anthropology,* Vol. 24, No. 2, pp. 269–290.

Qualter, T.H. (1970). *The electoral process in Canada.* Toronto : McGraw-Hill Ryerson.

Sancton, Andrew (1990). "Eroding representation-by-population in the Canadian House of Commons." *Canadian Journal of Political Science,* (September), pp. 441–457.

Tanguy, A. Brian (1994). "The transformation of Canada's party system in the 1990s." In Bickerton, James P. and Alain-G. Gagnon (eds.). *Canadian politics. 2nd ed.* Peterborough, Ont. : Broadview Press.

Thomas, Paul (1991). "Parties and regional representation." In Bakvis, Herman (ed.). *Representation, integration, and political parties in Canada, Vol. 14 of the research studies of the Royal Commission on Electoral Reform and Party Financing.* Ottawa : Privy Council Office.

Thorburn, H.G. (ed.) (1996). *Party politics in Canada. 7th ed.* Scarborough, Ont. : Prentice-Hall Canada.

Winn, Conrad and John McMenemy (1976). *Political parties in Canada.* Toronto : McGraw-Hill Ryerson.

Young, Lisa (2000). *Feminists and party politics.* Vancouver : University of British Columbia Press.

Liberal Party of Canada

Goldfarb, Martin and Thomas Axworthy (1988). *Marching to a different drummer.* Toronto : Stoddart.

McCall-Newman, Christina (1982). *Grits: an intimate portrait of the Liberal Party.* Toronto : Macmillan.

Wearing, Joseph (1981). *The L-shaped party: the Liberal Party of Canada, 1958–1980.* Toronto : McGraw-Hill Ryerson.

Whitaker, Reginald (1977). *The government party: organizing and financing the Liberal Party of Canada, 1930–1958.* Toronto : University of Toronto Press.

Bloc Québécois

Cornellier, Manon (1995). *The Bloc.* Translated by Robert Chodos, Simon Horn, and Wanda Taylor. Toronto : J. Lorimer.

Progressive Conservative Party

English, John (1977). *The decline of politics: the Conservatives and the party system: 1901–1920.* Toronto : University of Toronto Press.

Granatstein, J.L. (1967). *The politics of survival: the Conservative Party of Canada, 1939–1945.* Toronto : University of Toronto Press.

Martin, Patrick, Allan Greg, and George Perlin (1983). *Contenders: the Tory quest for power.* Scarborough, Ont. : Prentice-Hall Canada.

Perlin, George (1980). *The Tory syndrome: leadership politics in the Progressive Conservative Party.* Montreal : McGill-Queen's University Press.

Simpson, Jeffrey (1980). *Discipline of power: the Conservative interlude and the Liberal restoration.* Toronto : Personal Library.

Taylor, Charles (1982). *Radical Tories.* Toronto : House of Anansi.

Canadian Co-operative Federation/New Democratic Party

Avakumovic, Ivan (1978). *Socialism in Canada*. Toronto : McClelland and Stewart.

Lipset, S.M. (1950). *Agrarian socialism: the Cooperative Commonwealth Federation in Saskatchewan*. Los Angeles : University of California Press.

McDonald, Lynn (1987). *The party that changed Canada: the New Democratic Party, then and now*. Toronto : McClelland and Stewart.

Morton, Desmond (1986). *The new democrats, 1961–1986: the politics of change*. Toronto : Copp Clark Pitman.

Whitehorn, Alan (1992). *Canadian socialism: essays on the CCF and the NDP*. Toronto : Oxford University Press.

Young, Walter (1969). *Anatomy of a party: the national CCF, 1932–1961*. Toronto : University of Toronto Press.

Reform Party/Canadian Alliance

Dabbs, Frank (1997). *Preston Manning: the roots of reform*. Vancouver : Greystone Books.

Dobbin, Murray (1991). *Preston Manning and the Reform Party*. Toronto : J. Lorimer.

Flanagan, Thomas (1995). *Waiting for the wave: the Reform Party and Preston Manning*. Toronto : Stoddart.

Grubel, Herbert G. (2000). *A professor in Parliament: experiencing a turbulent Parliament and Reform Party caucus, 1993–97*. Vancouver : Gordon Soules Book Publishers.

Harrison, Trevor (1995). *Of passionate intensity: right-wing populism and the Reform Party of Canada*. Toronto : University of Toronto Press.

———— (2002). *Requiem for a heavyweight: Stockwell Day and image politics*. Montreal : Black Rose Books.

Laycock, David (2002). *The new right and democracy in Canada: understanding Reform and the Canadian Alliance*. Don Mills, Ont. : Oxford University Press.

Melnyk, George (1993). *Beyond alienation: political essays on the West*. Calgary : Detselig Enterprises.

Sharpe, Sydney and Don Braid (1992). *Storming Babylon: Preston Manning and the rise of the Reform Party*. Toronto : Key Porter.

Social Credit

Finkel, Alvin (1989). *The social credit phenomenon in Alberta*. Toronto : University of Toronto Press.

Irving, J.A. (1959). *The social credit movement in Alberta*. Toronto : University of Toronto Press.

Macpherson, C.B. (1953). *Democracy in Alberta: social credit and the party system*. Toronto : University of Toronto Press.

Pinard, Maurice (1971). *The rise of a third party: a study in crisis politics*. Scarborough, Ont. : Prentice-Hall Canada.

Stein, Michael (1973). *The dynamics of right-wing politics: a political analysis of social credit in Quebec*. Toronto : University of Toronto Press.

WEBLINKS

The Study of Canadian Political Party Member

This study, conducted under the auspices of Mount Allison University and supported by the Social Science and Humanities Research Council of Canada, is the first comprehensive study of political party membership in Canada. The study was designed to extend the understanding of changes to the Canadian party system through an exploration of the dynamics of party membership. The study set out the following categories of research questions:

1. What role do members play in the five federal parties? What motivates party membership and activity?
2. How do party activists view issues concerning representation and democracy?
3. Are there significant variations in the ideological orientations and attitudes toward policy issues among members of the various parties? How do these attitudes and divisions relate to the attitudes and patterns within the electorate?
4. What are the demographic profiles of each party membership? Do these profiles resemble those of the party's electoral supporters?

To generate feedback on these questions, the researchers distributed a total of 10,928 surveys to party members of each of the five major federal parties (i.e., Liberals, Conservatives, New Democrats, Canadian Alliance, and Bloc Québécois) in the spring of 2000. The researchers received 3,872 completed surveys upon which to base their findings. This represented an overall response rate of 36%. See **www.mta.ca/faculty/socsci/polisci/scppm/index.html.**

Registered Political Parties:

Bloc Québécois

www.blocquebecois.org

Canadian Action Party

www.canadianactionparty.ca

Canadian Reform Conservative Alliance

www.canadianalliance.ca

Communist Party of Canada

www.communist-party.ca

Green Party of Canada

www.green.ca

Liberal Party of Canada

www.liberal.ca

Marijuana Party

www.marijuanaparty.com

Marxist-Leninist Party of Canada

www.cpcml.ca

Natural Law Party of the United States

www.natural-law.org

New Democratic Party

www.ndp.ca

Progressive Conservative Party of Canada

www.pcparty.ca

QUESTIONS FOR CONSIDERATION AND DISCUSSION

1. What is the best method of choosing a party leader?

    _____

    _____

    _____

    _____

    _____

    _____

    _____

2. What are the ideological differences among the various political parties in Canada?

    _____

    _____

    _____

    _____

    _____

    _____

    _____

    _____

    _____

    _____

    _____

    _____

3. Is it appropriate for police organizations to provide ongoing security protection to elected officials during their re-election campaigns?

_____

_____

_____

_____

_____

_____

_____

_____

4. Are special interest groups more effective than formal political parties in addressing important policy issues in Canada?

_____

_____

_____

_____

_____

_____

_____

_____

_____

_____

_____

_____

_____

_____

5. What has caused the simultaneous appearance of five major parties in Canadian federal politics?

_____

_____

_____

_____

_____

_____

_____

_____

6. Which parties best reflect the policy interests of Canada's police organizations?

_____

_____

_____

_____

_____

_____

_____

_____

_____

_____

RELATED ACTIVITIES

1. Examine the website of one of the political parties in Canada and consider its position on issues pertaining to law enforcement, policing, or public safety.

2. Review the material referenced in the "Weblinks" section in this chapter dealing with the Study of Canadian Political Party Members conducted in 2000 by Mount Allison University. Discuss the details of the survey and consider the relevance of this project for your studies.

3. Acquire *Exploring Canada's Electoral System,* a CD-ROM produced by Elections Canada, and review the multi-faceted contents of this learning tool containing video clips, notebooks, quizzes, and games.

4. Review the following videotapes:

National Film Board of Canada (1978). *Reflections on a leadership.* Ottawa : National Film Board of Canada.

This is a fascinating study of the first bid by a woman for the leadership of the Progressive Conservative Party. She ultimately lost to Joe Clark; however, this is a documentary that shows how the party leadership process works in Canada. It is a worthwhile study that includes rare footage of the actual convention and valuable commentary from leading journalists, political scientists, and others on the event.

National Film Board of Canada (1979). *History on the run: media and the '79 election.* Ottawa : National Film Board of Canada.

This film provides an instructive and candid examination of the growing influence of the media in political campaigns. It looks at the 1979 federal election that resulted in the election of Joe Clark and the Progressive Conservative Party of Canada to a minority government. This election was a battle that included Pierre Elliott Trudeau, Joe Clark, and Ed Broadbent, and the film contains some extremely interesting footage of behind-the-scenes negotiations and discussions.

chapter five

Policing in a Political Context

So we must choose from among our Guardians those who appear to us on observation to be most likely to devote their lives to doing what they judge to be in the interest of the community [sic], and who are never prepared to act against it.

—Plato, *The Republic*, Part IV, 412 d-e (translated by Desmond Lee)

Royal Newfoundland Constabulary.

Contingent of Royal Newfoundland Constabulary officers en route to the Badger Strike, 1958.

Learning Objectives

1. Identify the purpose and key characteristics of special interest, or pressure, groups in Canadian politics.
2. Describe the ways in which lobbying efforts influence public safety and policing issues in Canada.
3. Outline several ways in which public opinion is gathered in Canada.
4. Cite several examples of political controversy that have involved matters of public safety or policing in Canada.
5. Outline the influence of the mass media on public safety and policing in the Canadian context.

INTRODUCTION

This chapter will look at how policing operates within the political context of Canada. Such an examination will involve a consideration of the aspects of the political realm that affect the realities of policing. One of the most important realities for modern policing is the presence, and influence, of special interest groups. Also referred to in the literature as "pressure groups," they frequently function to bring specific policy issues to the attention of police organizations and the nation's lawmakers. As organized entities, they are extremely important to our understanding of policing in a political milieu.

Next, this chapter will move to an overview of the whole process of lobbying on public safety issues. Lobbying, very simply stated, is the *modus operandi* of any special interest group. While lobbying in the strictest sense may take place even on an individual basis, it is the well-organized, well-orchestrated, well-funded efforts of professional lobby groups that typically influence public policy-making in the realm of policing. This chapter will focus on several key lobby groups and will outline some of their most relevant policy issues.

This chapter also takes a look at the issue of public opinion and its role in the public safety realm. There are several key organizations that undertake public opinion polling in Canada. These organizations will be identified and their efforts will be briefly summarized.

The next topic addressed relates to political controversy in the area of policing and public safety. Several important instances will be reviewed to demonstrate that the policing portfolio is one that can be highly volatile for politicians and bureaucrats alike.

Finally, this chapter will look at the role of the mass media in relation to policing matters. Because policing and public safety issues hold a strong fascination for the public, and because they are typically closely tied to at least one of the three levels of government, the Canadian mass media are virtually unrelenting in their fixation with such matters.

SPECIAL INTEREST GROUPS

The term "special interest groups" refers to the concentration of organizations and agencies with a strongly voiced concern regarding some particular aspect of a public issue. Our attention will focus on special interest groups with a substantial interest in issues of policing or public safety in Canada. Given the importance of policing within our society and the special nature of the powers that are bestowed on police officers, it is not surprising that a number of special interest groups have come into being around such issues. The following listing presents some of the categories of groups that have a sustained interest in Canadian policing and public safety:

- race relations groups;
- women's rights groups;
- gun enthusiast groups;
- gay and lesbian groups;
- freedom of information and protection of privacy groups;
- judicial groups;
- defence counsel and Crown attorneys;
- prisoners' rights groups;

- mental health groups;

- civil liberties and human rights groups;

- child and youth advocates;

- drug law reform groups; and

- victims' rights groups.

Each of these groups may have a particular and well-defined interest in policing and public safety. Many such groups wish to pursue a specific policy direction and make their position known to the police organization(s) or government within their jurisdiction. Special interest groups have existed for a long time and, in many instances, have had a profound impact on the development and implementation of public safety and policing policy in Canada.

White, Wagenberg, and Nelson (1994) have observed that many special interest groups have formalized their status and possess a high level of organization:

> Some major interest groups in Canada have established themselves as federations in order to influence the local, provincial, and federal governments within the Canadian federal structure. This has resulted in many of them having head offices in Ottawa and offices in various provincial capitals, all of which are staffed by permanent employees. *(p. 145)*

Interest groups may also be referred to as "pressure groups." Clearly, one of the key tactics of any organized group that wishes to influence public policy is to bring pressure to bear on the pertinent decision-makers in ways that promote the group's policy agenda. Many interest groups have become highly adept at applying the right kind of pressure on public policy decision-makers.

To sustain a high level of interest and cohesion, pressure groups must work at developing strategies that will engage and expand their membership. Accordingly, the most efficient and effective special interest groups spend a great deal of time conducting conferences, workshops, and seminars that seek to explore key topics and policy options. Many groups will invite keynote speakers to address their concerns and assist them, not only with the challenges of securing the support of the decision-makers, but also with becoming more effective lobbyists. These groups often have a regular newsletter or journal that is issued to members to sustain awareness and interest, as well as to promote discussion within the group.

Special interest groups in Canada often expend a great deal of effort on their own public relations campaigns. It is perhaps useful to emphasize that police organizations themselves are frequently engaged in activities that reflect their own "special interests." In the next section, we will consider a few relevant examples of police lobbying efforts.

LOBBYING ON PUBLIC SAFETY ISSUES

Lobbying, in our context, involves the efforts of any group that are intended to influence the outcome of policy deliberations relating to policing and public safety. While the term often has negative connotations, it is almost inevitable that organized groups with a specific perspective on some public policy issue will take it upon themselves to put considerable pressure on decision-makers to address their policy concerns in meaningful ways.

Lobby groups will operate on several fronts to ensure that their policy directions are realized to the highest degree possible. They will function to influence the parliamentarians and decision-makers at various levels of government with regard to their concerns. They will also work to educate the public to gain additional support for their policy directions and priorities. Furthermore, special interest groups will work quite closely with the bureaucratic layers of government to provide the detailed insight and support necessary in support of policy plans that will meet the complex demands of public safety and policing issues. Finally, lobby groups function to educate, mobilize, and consolidate support within their own organizations. It is obviously far more effective from an advocacy standpoint if any given lobby group can state that its membership is unanimously in favour of a particular policy direction. Clearly, to be effective, the work that a special interest group takes on requires a great deal of attention to detail and constant monitoring.

Pross (1990) has noted that governments find a great deal of comfort in knowing that special interest groups will offer their support for certain policy decisions. By addressing the key concerns of major pressure groups, governments stand to gain significant credibility with the public, who often view the special interest groups with more confidence than they view the governments themselves.

> [P]ressure groups play a very significant part in persuading both policy makers and the general public that changes in public policy are worthwhile, generally desired, and in the public interest. Because pressure groups frequently speak for a significant proportion of the public that will be affected by a change in policy, governments find it reassuring to have their proposals endorsed by the relevant groups. *(p. 287)*

In complex matters of policing and public safety, governments frequently prefer to enlist the competence, expertise, and capacity of the relevant professional groups to assist in the development of policies, programs, and procedures that will guide the regulation of these functions. Just as doctors, lawyers, teachers, and other professional groups have been accorded a high degree of autonomy in the regulation of their members, police organizations across Canada have been delegated a considerable margin of responsibility for the regulation of their profession (Pross, 1990, p. 289). This delegation of authority, of course, allows the government to hold itself somewhat aloof from potential controversy and criticism.

It is important to recognize that lobbying on public safety or policing issues is not something that rests exclusively with the kinds of special interest groups identified previously. Police organizations themselves are frequently at the forefront of policy matters that have an impact upon their members. A vast array of issues has been and will continue to be pursued by agencies, associations, and organizations affiliated with the policing community.

The concept of the "policy community" is worth examining in this context. Whenever there exists some matter of public policy that requires government attention, it is highly likely that a wide range of groups will be seeking to inform, influence, or otherwise affect the outcome of those policy deliberations. Figure 5.1 presents Dyck's (1993) summary of the possible range of elements within any "policy community."

Frequently, it will occur that the primary focus of a pressure group's activity will be on the governmental bureaucracy that has functional responsibility for the policy issues of concern to that group. Dyck (1993) explains the reason for this phenomenon as follows:

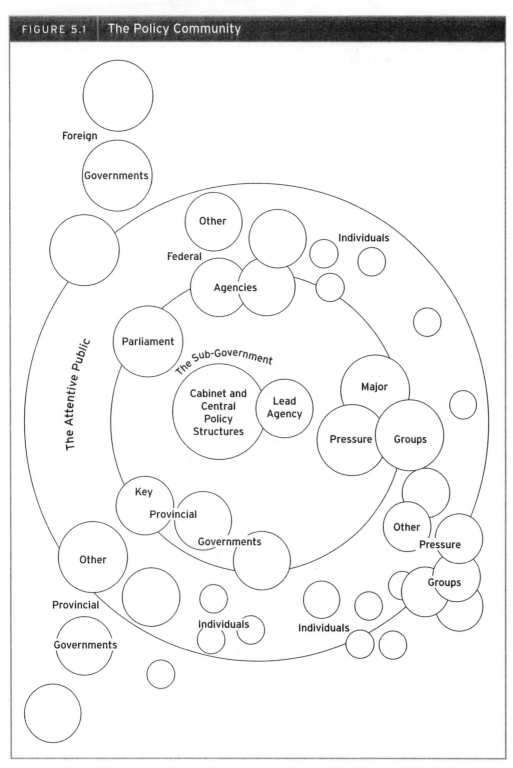

FIGURE 5.1 | The Policy Community

Source: Rand Dyck (1993). *Canadian politics: a critical approach.* Scarborough : Nelson Canada, p. 310. Used with permission.

Many pressure group demands involve technical matters that only the bureaucracy understands and that it may be able to satisfy without reference to the politicians. Although bureaucrats may initially view meeting pressure groups as a waste of precious time and be suspicious of their motives, such groups try to cultivate close relationships with the senior public servants so that they feel comfortable in contacting these officials on an informal, direct, day-to-day basis. *(p. 309)*

At the federal, provincial, and municipal levels, there are activities that require the detailed attention of special interest groups in the area of policing. Often this attention must also be extended to the bureaucratic layers that have a mandate to develop relevant policy options for approval by elected decision-makers. For example, in the late 1990s, it became necessary to introduce important changes in police legislation in Ontario. Accordingly, it fell to the following key groups, representing both bureaucratic and special interest groups, to assist the Solicitor General in the development of the legislative and regulatory framework that eventually took shape as the amended *Police Services Act:*

- Association of Municipalities of Ontario
- Ontario Association of Chiefs of Police
- Ontario Association of Police Services Boards
- Police Association of Ontario
- Ontario Provincial Police Association
- Ontario Ministry of the Attorney General
- Ontario Provincial Police
- Policing Services Division, Ministry of the Solicitor General
- Ontario Senior Police Officers' Association.

It should also be noted that there is often a considerable degree of employee movement from the special interest groups to the government bureaucracy and vice versa. Due to the specialized knowledge, skills, and abilities that are required to function at a high level of competence in the areas of policing and public safety, it is not uncommon for senior police managers to transfer into the bureaucracy as public service executives. Police executives in Ontario have been recruited at senior levels within the Ministry of the Solicitor General, and conversely, senior public servants have been placed at the highest levels of police departments within the province.

In Canada, many interest groups have dealt with policing and public safety issues as part of their overall mission and mandate. The following listing presents a sample of the key groups that have taken an interest in policing and public safety issues; and while there may be many other lobby groups that tackle similar issues, this listing provides a good overview of the most relevant federations:

- Amnesty International
- Assembly of First Nations
- Canadian Association of Chiefs of Police
- Canadian Association of Civilian Oversight of Law Enforcement
- Canadian Association of Police Boards
- Canadian Bar Association
- Canadian Civil Liberties Association
- Canadian Police Association
- Canadian Teachers' Federation

- Federation of Canadian Municipalities

- John Howard Society

- Mothers Against Drunk Driving Canada

- Ontario Association of Chiefs of Police

- Ontario Association of Police Services Boards

- Police Association of Ontario

Amnesty International

This group, which is part of a global network of affiliated chapters, takes a keen interest in activities that pertain to human rights and the treatment of political prisoners. Each year Amnesty International releases a detailed report that highlights issues of concern in various countries where there are breaches of human rights and instances of torture and punishment of political prisoners and others. In 2002, Amnesty International's report for the period January to December 2001 targeted Canada as a jurisdiction where the following issues were of serious concern:

- Police use of excessive force in controlling demonstrators at the Summit of the Americas held in Quebec City. In Amnesty International's estimation, the use of tear gas and plastic bullets in this instance warrants the convening of an independent inquiry into allegations of police brutality.

- The beating death of Mr. Otto Vass by four officers of the Toronto Police Service in August 2000.

- Incidents involving First Nations peoples and members of the Saskatoon Police Service. Allegations concerning the mistreatment of First Nations peoples by the department prompted an inquiry into these matters. Of particular concern were the deaths of two men, Rodney Naistus and Lawrence Wegner, who were found frozen on the outskirts of the city in early 2000.

- The absence of an independent inquiry into the shooting death of Dudley George by an Ontario Provincial Police officer during a First Nations land claim protest at Ipperwash Provincial Park.

Assembly of First Nations

The Assembly of First Nations (AFN) represents many of Canada's Aboriginal peoples. The AFN has been clear in its determination to establish a separate justice system for First Nations peoples in Canada. As a result, the AFN sees its relationship with the federal government as being a "nation-to-nation" dialogue. As part of its organizational structure, the AFN has a Justice Secretariat that pursues strategic objectives relevant to policing and public safety. One of the key areas of activity for the Justice Secretariat is the relationship of the AFN with the RCMP and any matters that pertain to Aboriginal policing across Canada.

Canadian Association of Chiefs of Police

This is an essential organization that has had an enormous impact on the development of policing legislation and policy in Canada. The Canadian Association of Chiefs of Police (CACP) began in Toronto in 1905 as the Chief Constables Association of Canada. It is dedicated to facilitating progressive change in policing in Canada. One of its more recent initiatives was the creation of the Police Futures Group, a think-tank designed to examine critical future issues in policing and to foster dialogue, debate, and discussion on those issues. Among the overall objectives of the CACP are the advocacy of

- legislative reform, resource allocation, and policy improvements with the people of Canada and their governments;
- innovative solutions for crime and public order issues with CACP partners and concerned people of Canada;
- community partnerships with the people of Canada; and
- the highest professional and ethical standards within the police community.

Every year, during its annual conference, the CACP passes a series of resolutions that frequently pertain to legislative or regulatory provisions in the area of Canadian justice. Since the CACP is a significant stakeholder in the area of policing and public safety, its positions are taken seriously not only by Canadian lawmakers, but also by the Canadian public.

Canadian Association of Civilian Oversight of Law Enforcement

The Canadian Association of Civilian Oversight of Law Enforcement (CACOLE) is a national organization of various agencies tasked with responsibility for the oversight of police services in Canada. The association represents police commissions, police services boards, and other bodies and tribunals that advance the principles and practices of civilian oversight of law enforcement. CACOLE produces a regular newsletter and hosts an annual conference that features sessions of interest and relevance to its membership.

Canadian Association of Police Boards

The Canadian Association of Police Boards (CAPB) is an organization that specializes in promoting the delivery of effective governance for policing across Canada. It works in partnership with the federal government, particularly the Minister of Justice and the Solicitor General, as well as provincial and local agencies, in pursuit of safer and more secure communities. The CAPB was formed in 1989 and is currently based in Ottawa. It works toward formulating a cohesive perspective reflecting the views of civilian governing authorities on policy issues affecting police organizations. Furthermore, the CAPB acts as a liaison linking federal, provincial, and municipal government authorities and the federal and provincial Solicitors General.

In its advocacy role, the CAPB has made, and continues to make, representations to the federal government on behalf of its members. Some of the areas that the CAPB has provided detailed input on include

- gun control;

- DNA testing;

- amendments to the *Young Offenders Act;*

- child prostitution;

- review of the national police services;

- criminal liability; and

- anti-terrorism.

Early each year, the board of directors of the CAPB meets with the Minister of Justice and the Solicitor General of Canada. This affords the CAPB an opportunity to highlight significant emerging issues and to present its position on those issues. Conversely, it provides the federal ministers with an invaluable forum for consulting with key policing stakeholders and seeking their perspective on government initiatives within the fields of justice, policing, and public safety.

The CAPB holds an annual conference where major concerns are addressed and resolutions passed for presentation to the appropriate government agencies. Additionally, the CAPB has collaborated with other special interest groups to offer seminars and workshops on critical issues affecting policing and public safety in Canada. In November 2002, the CAPB and the Ontario Association of Police Services Boards co-hosted a seminar on the future roles of public police and private security agencies. This program coincides with the release of a discussion paper, prepared under the auspices of the Law Commission of Canada, entitled *In search of security: the roles of public police and private agencies.*

Canadian Bar Association

The Canadian Bar Association (CBA) was formed in 1896 and currently represents the professional interests of more than 37,000 lawyers, judges, notaries, law teachers, and law students in Canada. The CBA advocates on behalf of its membership and provides them with personal and professional support, including professional development programs and publications.

The CBA works to facilitate meaningful and appropriate law reform and promotes equality within the Canadian legal profession. Specifically, the mandate of the CBA is to

- improve the administration of justice;

- improve and promote the knowledge, skills, ethical standards, and well-being of members of the legal profession;

- represent the legal profession on a national and international level;

- promote the interests of the members of the CBA; and

- promote equality in the profession.

With respect to law reform, the CBA maintains an active involvement in the development of government policy. It undertakes lobbying efforts on issues that have been identified by its members and will respond to government initiatives that touch upon relevant concerns of the organization. The CBA presents its views on policy and legal matters by means of written submissions, appearances before parliamentary committees, informal meetings with the federal Minister of Justice, and regular consultations between various CBA sections and the Department of Justice or other pertinent government officials. Furthermore, the CBA frequently intervenes at the Supreme Court of Canada on matters that hold a compelling public interest or are of special importance to the Canadian legal profession.

Canadian Civil Liberties Association

The Canadian Civil Liberties Association (CCLA) is a lobbying and law reform organization that maintains a focus on issues of civil liberties and human rights. It is a non-profit association that accepts no government funding and is spearheaded by A. Alan Borovoy.

An exceptionally active organization that conducts studies on issues affecting the human rights and freedoms of Canadians, the CCLA has been extremely vocal about police matters. It has issued several detailed media releases addressing matters of concern with respect to the police community. What follows is a brief sample of the topics that have been tackled by the CCLA:

- Bill C-35, which empowers the Royal Canadian Mounted Police (RCMP) to introduce security measures it deems "reasonable" during international gatherings such as the G-8 meeting in Kananaskis, Alberta;

- the creation of independent audit agencies in every jurisdiction in Canada with authority to access police records, facilities, and personnel (working in a manner similar to the Security Intelligence Review Committee, which oversees the Canadian Security Intelligence Service);

- specific guidelines for police strip searches; and

- political fundraising by the Toronto Police Association.

Canadian Police Association

This umbrella police union association has become extremely well organized and on behalf of its membership presents to the federal government an incredible range of briefs and resolutions pertaining to policing and public safety issues. The following listing offers a small sample of its various briefs and other submissions relating directly to justice reform in Canada:

- presentation to the Standing Committee on Justice and Human Rights regarding the statutory review of the mental disorder provisions of the *Criminal Code*;

- brief on organized crime;

- brief on Bill C-22, the *Proceeds of Crime (Money Laundering) Act;*

- brief on Bill C-3, the *DNA Identification Act,* and consequential amendments to the *Criminal Code* and other acts;

- brief on Bill C-16 (regarding Powers to Arrest and Enter Dwellings);

- brief on Bill C-27 (an act to amend the *Criminal Code,* concerning child prostitution, child sex tourism, criminal harassment, and female genital mutilation);

- brief on Bill C-37 (an act to amend the *Young Offenders Act* and the *Criminal Code*).

The Canadian Police Association (CPA) recognizes that to become an effective lobby group, it must present itself in ways that are both professional and persuasive. Working toward the accomplishment of its goals, this association has adopted a strategic approach to advocacy. Accordingly, the CPA is prepared to use public opinion polling and other tools that are employed by political parties to ensure that its policy objectives are acceptable to a wider audience beyond its own membership. As noted by Griffin (2002):

> Advocacy is often the single most important function for a wide range of associations, grass root organizations, lobbyists and special interest groups. From chicken farmers to drug companies, victims of crime to competitive sports organizations, associations representing a common interest want to influence public policy in their sectors. *(p. 12)*

It is essential that associations know what they're doing when they engage the powers-that-be in their efforts to guide policy formulation and implementation at whatever level. They need to be aware of political party platforms and have access to those who are responsible for developing those platforms in the area of concern for the association. They need to foster and cultivate relationships with the decision-makers who will construct public policy, and they need to establish strategic alliances that will assist them in the presentation of their issues to the policy-makers in government.

Since 1995 the CPA has held a National Lobby Day, during which its representatives meet with politicians on Parliament Hill to engage in a dialogue on policing and public safety issues. The CPA represents more than 29,000 members, and this event provides a forum for parliamentarians to meet with spokespersons who are able to address front-line issues of policing, justice, and health and safety. In 2002, CPA delegates concentrated their discussions on four issues:

- protection for police officers and victims of crime;
- lenient conditions in federal penitentiaries;
- protection of children; and
- police funding.

Canadian Teachers' Federation

The Canadian Teachers' Federation (CTF) was founded in 1920 to represent the interests of teachers across the country. A national umbrella organization with 14 provincial and territorial members, it speaks on behalf of more than 240,000 teachers in Canada. As an advocacy group, the CTF works at the federal level on a wide range of issues of significance to teachers and to public education in general. In the area of policing and public safety issues, the CTF is particularly active when it comes to young offenders and youth violence. This organization is deeply concerned about the possible links between education and crime prevention. It is committed to establishing realistic expectations within the schools and to developing new partnerships that will work to reduce youth crime and violence. One specific focus for the CTF relates to section 43 of the *Criminal Code,* which provides a defence for teachers and others charged with assault against children in their care. There have been arguments made that this section violates the *Canadian Charter of Rights and Freedoms,* and the CTF is seeking to support this section to maintain safe and secure learning environments that will, additionally, protect teachers against unnecessary liability.

Federation of Canadian Municipalities

As its name implies, the Federation of Canadian Municipalities (FCM) is a national association representing the concerns and priorities of the municipal corporations across Canada. It is made up of mayors and elected municipal councillors and pursues a number of issues that are related to policing and public safety. The FCM is an extremely important organization that holds a great deal of power when it comes to policing and public safety issues in Canada. Of particular relevance for the FCM are matters pertaining to community safety and crime prevention. This organization is dedicated to supporting measures that will contribute to the maintenance and enrichment of safe communities. With approximately $5.7 billion being spent on policing services across Canada annually, the FCM is determined to work in cooperation with provincial and federal authorities to ensure that these costs are reduced and steps are taken to enhance the quality of life in Canada's cities.

During its annual conference in 2002, the FCM adopted a detailed policy statement on community safety and crime prevention. This statement builds upon the federation's earlier national philosophy on the creation of stronger, safer community life in Canada and reaffirms its commitment to inter-agency cooperation in the prevention of crime. The FCM recognizes that reactive policing services are not adequate for the challenges of public safety and security. It is essential that there be more proactive efforts taken to address the causes of crime and disorder within society. The FCM has committed itself to working in partnership with national and international organizations toward the goal of developing and sustaining safer communities. This includes a close affiliation with the National Crime Prevention Centre in Ottawa.

Among its many strategies for addressing the issues of crime prevention and community safety, the FCM has articulated the following elements that pertain specifically to policing:

- foster police support for citizen and community agency crime prevention efforts;

- encourage police agencies to work more in the community to prevent crime and reduce fear, particularly through problem-oriented policing approaches;

- ensure that adequate programs and inter-agency approaches are available for those already convicted of crime;

- provide better information, access to remedies, and reparation for victims of crime;

- encourage municipal leaders to ensure greater use of mediation and conciliation as a response to criminal conflict;

- encourage municipal leaders to provide their staff, police, and community leaders with the appropriate training on inter-agency approaches to crime prevention; and

- ensure that police personnel receive training to assist them in dealing sensitively with racial and ethnic minorities. *(Adopted June 2001, FCM annual conference, revised March 2002, Standing Committee on Community Safety and Crime Prevention)*

Furthermore, the FCM has adopted several resolutions that pertain to major policing and public safety issues in Canada. Of particular concern to the FCM are the challenges presented by organized crime. In 2000, the FCM hosted a major national symposium designed to promote a community response to organized crime. This symposium, supported by the federal Solicitor General, was held in cooperation with the Criminal

Intelligence Service of Canada, the Department of Citizenship and Immigration Canada, the Department of Finance Canada, the Department of Justice Canada, the Montreal Urban Community, and the RCMP.

John Howard Society

This organization was started by a small group of church workers in Toronto in 1867. Its goal was to bring spiritual assistance to prisoners in the local jail. Currently, the John Howard Society is a federation of local and provincial societies working to find appropriate and humane responses to the root causes and consequences of crime in society. With offices across Canada, the society develops briefs and position papers on policy issues that pertain to their mission. Recently, the society has prepared research documents dealing with literacy, correctional reform, and legislation on young offenders.

Mothers Against Drunk Driving Canada

This organization has taken as its mission the eradication of impaired driving and seeks to provide support to the victims of this violent crime. In pursuit of that mission, Mothers Against Drunk Driving (MADD) Canada pursues a vigorous public awareness campaign aimed at raising understanding of the dangers of impaired driving. The organization is made up of a national body and several local chapters. One of its programs is an annual award of excellence in police services, which recognizes important contributions made by law enforcement agencies across the country in the reduction of impaired driving. In 2002, the award was renamed in honour of Terry Ryan, a popular officer with the Durham Regional Police Service who was killed by an impaired driver in May of that year.

As part of its efforts to lobby the government for stricter legislative measures against impaired driving, MADD Canada sponsored a national survey of 2,000 Canadians that revealed the following public support:

- nine out of ten Canadians (i.e., 90 percent) agree that driving under the influence of alcohol is a significant safety problem on our highways;

- fifty-five percent of Canadians believe that both federal and provincial governments need to do more to reduce impaired driving; and

- nearly eight out of ten Canadians (i.e., 77 percent) believe that impaired driving represents a major problem for young people aged 16 to 20 years.

Accordingly, MADD Canada strongly supports legislative measures to increase enforcement powers for the police with respect to impaired driving.

Ontario Association of Chiefs of Police

This provincial association has a strong attachment to policing and public safety issues. As part of its regular activities, the Ontario Association of Chiefs of Police (OACP) assumes an advocacy role that includes defining significant policing issues and developing clear and conscientious positions to promote in partnership with the community.

In 2002, the executive of this association met with the Minister of Public Safety and Security to brief the minister on the following key issues that were important to the OACP:

- *Police restructuring process.* The OACP sees an urgent need for a comprehensive review of police restructuring in Ontario over the last five years in light of the reduction in the number of municipal police services from 108 in 1996 to 69 today.

- *Domestic Violence Protection Act (Bill 114).* The OACP takes the position that this legislation should be deferred until a long-term, sustainable model for the effective response to domestic violence can be arrived at.

- *Justices of the peace and video remand.* The OACP takes the position that there should be an increase in the number of justices of the peace available to meet the demands of police services, and the video remand project should be put into place immediately.

- *Centralized training for Ontario police.* The OACP supports a centralized training model and seeks to ensure that police services retain control of the training standards and methods.

Recently, the OACP launched an aggressive government lobbying strategy to increase its presence with provincial government authorities and fortify its capacity to achieve effective results with respect to key policing issues. As part of that strategy, the OACP has entered into a partnership with a Toronto-based government relations firm.

Ontario Association of Police Services Boards

In 1962 an organization known as the Municipal Police Authorities was formed to deal with civilian police governance in Ontario. Today, that group's direct descendant is the Ontario Association of Police Services Boards (OAPSB), a non-profit organization that offers a range of services to its members to assist them in meeting their oversight and governance responsibilities for policing services.

The OAPSB advocates on behalf of its members and deals with a comprehensive range of issues and matters that pertain to policing and public safety within the Province of Ontario. This entails engagement with government officials and elected members who have policy-making and legislative responsibility within the policing portfolio. Also, the OAPSB acts on behalf of its various board members as a clearing house for information from government.

The OAPSB frequently collaborates on major policing and public safety issues with its national counterpart, the Canadian Association of Police Boards, and also works in close cooperation with the Ontario Association of Chiefs of Police and the Police Association of Ontario. The OAPSB maintains an up-to-date website, hosts annual conferences for its members, and offers topical workshops, seminars, and training sessions for the continuous improvement of its members.

Police Association of Ontario

The Police Association of Ontario (PAO) is a powerful lobby in the province for matters pertaining to policing, law enforcement, and public safety. It represents more than 13,000 front-line municipal police officers and therefore constitutes a considerable lobby group that can wield a great deal of influence with the decision-makers responsible for policing policy and legislation. An example of the more recent engagement in the political process can be seen in Figure 5.2, which contains a media release distributed in August 2002 endorsing Mr. Paul Martin's leadership campaign for the federal Liberal Party of Canada.

FIGURE 5.2	PAO Endorsement of Paul Martin

POLICE ASSOCIATION OF ONTARIO

Media Release
For Immediate Release
August 16, 2002

Police Association of Ontario Endorses Paul Martin

(Windsor) – The President of the Police Association of Ontario (PAO), Bob Baltin, announced today that the membership of the PAO is officially endorsing the candidacy of Mr. Paul Martin for leader of the Federal Liberal Party. The membership of the PAO voted to endorse Mr. Martin at their 70th Annual General Meeting that is being held in Windsor. The PAO represents approximately 13,000 front-line municipal police personnel across the province.

"This country needs strong leadership and we believe that Mr. Martin will provide that. Mr. Martin has a strong track record of supporting initiatives to promote safer communities across Canada", stated Bob Baltin.

Bruce Miller, the Administrator of the PAO stated, "Front-line police personnel have called on the Federal Government to make needed changes to help ensure public safety. Our calls have fallen on deaf ears. We have not even had the courtesy of a reply. Mr. Martin has proven that he is concerned and is always willing to listen to any recommendation to make our streets safer for Canadians."

For more information contact:

Bob Baltin, PAO President
Bus: 905-670-9770

Bruce Miller, PAO Administrator
Bus: (905) 670-9770
Cell: (416) 707-3923

Source: Police Association of Ontario. Used with permission.

Often, the PAO will collaborate on important policing issues with other special interest groups to form a united front. The most common alliance occurs with the national body representing front-line police officers, the Canadian Police Association. However, increasingly, the interests of the PAO frequently coincide with the Ontario Association of Chiefs of Police, the Canadian Association of Chiefs of Police, and the Canadian Association of Police Boards.

An interesting example of a fundraising campaign conducted by a police union designed to generate monies for its lobbying efforts involves the Toronto Police Association (TPA). "Operation True Blue" was intended to be a public relations coup for an increasingly aggressive labour union representing front-line police officers in Toronto, under the presidency of Craig Bromell. The TPA hired a telemarketing firm to call Torontonians and solicit financial support for the association. As a result of a payment, contributors would receive a decal that could be displayed on their automobile windshield.

Reportedly, the campaign generated more than $306,000 in a mere 32 days, a majority of those funds going to the telemarketing firm. However, this campaign drew an enormous amount of anger from the Chief of Police, the Toronto Police Services Board, Toronto City Council, and the public. It became a focus for media attention during the early months of 2000 and caused many people to reconsider their support for the local police union. On their part, the executive of the Toronto Police Association argued that the "Operation True Blue" campaign was meant to allow them to pursue changes to the *Young Offenders Act* and other improvements in the Canadian criminal justice system. However, there was a considerable amount of skepticism with respect to this initiative, and there were fears that the police union was attempting to intimidate local politicians who had been critical of its members and policing in the city in general. Even the Canadian Civil Liberties Association weighed in on the controversy and demanded that the Chief of Police and the Toronto Police Services Board act to restrain the police union bosses, particularly Mr. Bromell.

PUBLIC OPINION

Public opinion polling has had a long-standing tradition in Canadian politics. It is an approach that has been highly influential in guiding the administration of public affairs. It has been used by many organizations, including police departments, to assist them in setting functional, operational, and strategic directions for their future. Political parties, government agencies, police associations, and many special interest groups have engaged the services of professional pollsters in an effort to guide their own policy formulation and to gauge public support for their preferred policy initiatives.

A few of the major public opinion polling firms operating in Canada today are described below.

EKOS

EKOS is a Canadian public opinion firm that has been in operation for more than two decades. It provides full-service consulting support in the following areas:

- market research;
- public opinion research;

- strategic communications advice;

- program evaluation;

- performance measurement; and

- organizational and human resources management research.

EKOS has prepared many studies and reports that have been published for public consumption, including a detailed study of public attitudes toward family violence (2002). Also, EKOS has recently inaugurated a syndicated research initiative designed to monitor public security initiatives.

Environics Research Group

This market research consulting business was started by Michael Adams. It has created several important tools such as the *Focus Canada Report,* a comprehensive survey of Canadian attitudes toward public policy issues including political, economic, and social trends, and the *Focus Ontario Report,* which presents a quarterly survey of Ontarians' attitudes on major issues in the province's agenda. The Ontario Provincial Police (OPP) used this organization to undertake an extensive survey of public attitudes toward the OPP in 1996.

Goldfarb Consultants

A major firm that does market research and public opinion polling, Goldfarb Consultants has served clients in both the public and private sectors. Martin Goldfarb, who founded the company in 1965, was for many years the official pollster for the Liberal Party. In 2002, Goldfarb Consultants merged with Millward Brown.

Hill & Knowlton Canada

This is another Canadian organization that provides services in the area of public relations, public affairs, and strategic communications. Hill & Knowlton Canada was created in 1989 and combines PAI, Canada's first government relations firm, Hill and Knowlton, a large communications organization, and Decima Research, an opinion research firm founded in 1979 by Allan Gregg.

Ipsos-Reid (formerly Angus Reid)

Another major Canadian marketing research and public opinion firm, Ipsos-Reid currently has a network of more than 3,000 researchers and a capacity to undertake extensive data collection and analysis on behalf of its clients.

Governments often feel a particularly strong need to ensure that there is substantial public support for policing and public safety initiatives. Polling offers a relatively straightforward, and momentarily reliable, approach to this challenge.

POLITICAL CONTROVERSY

Policing and public safety issues can be fraught with difficult challenges for politicians, police executives, and bureaucrats alike. It is only necessary to review the top headlines in most local newspapers to see that these issues are often behind the lead stories, and there is usually an avid interest in the outcome of such matters. It has been determined that high-profile policing issues are a prime source of stress for police executives (Biro et al., 2000), and they are also a major source of controversy for politicians and public servants at all levels and across all jurisdictions. What follows are three examples of politicians who have gotten into serious difficulties as a result of their attachment to the policing or public safety portfolio.

The APEC Inquiry and Solicitor General Andy Scott

In 1997 the City of Vancouver played host to the Asia-Pacific Economic Cooperation (APEC) summit. Prime Minister Chrétien was eager to have a successful and peaceful gathering of international delegates. Accordingly, members of the RCMP were joined by local law enforcement agencies to provide security and crowd management. Unfortunately, the RCMP took actions in the area of crowd control that have been widely criticized, including the use of pepper spray on student demonstrators and camera crews. As a result of these police tactics, the RCMP Complaints Commission began an inquiry into policing at the APEC summit in October 1998. As the inquiry hearings were progressing, Andy Scott, the federal Solicitor General at the time, was overheard telling a friend details about the likely outcome of the inquiry during an airplane trip to his home constituency. Unfortunately for Andy Scott, the person who overheard these comments was Dick Proctor, an NDP Member of Parliament. The political pressure, and the intense media scrutiny, continued to mount until the Solicitor General's position was no longer tenable. On November 23, 1998, he tendered his resignation from the Cabinet.

Lawrence MacAulay and Government Contracts

In 2002 considerable public concern was aroused by the involvement of the federal Solicitor General, Lawrence MacAulay, in the awarding of $14.7 million to Holland College in Prince Edward Island. This community college, which is headed by the Solicitor General's brother Alex, received a $6.5 million grant under the Atlantic Canada Opportunities Agency for research and development associated with the creation of Internet-based training courses for police and corrections officers. In addition, both the RCMP and Corrections Canada had turned down grant applications from the college until MacAulay intervened on the school's behalf. In the House of Commons, the Solicitor General was criticized for lobbying the RCMP Commissioner for approval of a $3.5 million grant application. More recently, MacAulay has come under scrutiny for the award of a contract worth more than $100,000 to a friend's accounting firm. What has puzzled government critics is the Solicitor General's argument that his friend's accounting firm is capable of providing his department advice and assistance in matters pertaining to national security and the justice system. These matters have raised concerns about the ethical standards within the Liberal government and have provoked much criticism of the Prime Minister.

Mike Harris and the Ipperwash Protest

On September 6, 1995, Dudley George was shot and killed by an OPP officer during an occupation of Ipperwash Provincial Park in southwestern Ontario. Sergeant Kenneth Deane, the OPP officer responsible for the shooting, was convicted of criminal negligence causing death. However, serious issues were raised about the possible involvement of Premier Mike Harris's office in the ultimate decision to use force during this protest. Over the last several years, there have been many calls for an independent inquiry into this matter. The United Nations Human Rights Committee has urged the Canadian government to call an inquiry, and Amnesty International has established a campaign to urge the Prime Minister to take action on this matter. While no politician or civil servant has been removed for involvement in this affair, excluding Sergeant Deane, there continues to be an aura of controversy around this occurrence that may still have an impact on the provincial government.

MASS MEDIA

The influence of mass media on policing is profound. We are surrounded by media messages coming at us from newspapers, magazines, radio, television, and even outdoor displays. The electronic world envisioned by Marshall McLuhan has come to pass, and we truly exist within a "global village" where news of events in distant countries comes to us almost instantaneously. This barrage of media has had an impact on the way governments do business. Likewise, the infinite demands of an extensive media industry have made it important for politicians, public servants, and police leaders to understand how to manage those demands. Dyck (1993) has noted the incredible importance of the mass media within the Canadian political system:

> It is now commonly observed that the mass media and public opinion polls are two of the most important elements in the political system…[They] predominate in performing the function of "political communication" between public and government and among different branches of government. In addition, few of the other political functions could be performed without at least partial dependence on the media. *(p. 205)*

Over the last 10 years there has been a considerable rise in the number of media relations courses available for police managers and executives. It has become quite clear that police administrators must have the knowledge, skills, and abilities necessary to deal with an array of media relations issues, and they must do so in a competent, conscientious, and confident manner. A recent study conducted on behalf of the Canadian Association of Chiefs of Police made it clear that the news media can be a source of anxiety for many police executives (Biro et al., 2000). However, there is no value in avoiding the media. Modern police organizations must learn to capitalize on the opportunities to present an effective message to the public through the various media outlets currently available.

Increasingly, police organizations have come to recognize that in fact there are ways to engage the media as a tool for functioning within a complex political environment and for disseminating important information and positions to the public. Many progressive police organizations in Canada are now fully aware of the key elements of a media or communications strategy. There has been a move to employ approaches and techniques that have been learned by virtually every politician and senior bureaucrat to "sell" their programs and policies. By working to construct and manage their own issues and messages, police organizations can work with the mass media in the dissemination of information to the public.

If one considers the complexity of media issues at the level of the Prime Minister's Office, it quickly becomes clear that the role of communications is central to political success. There are dozens of staff members whose exclusive function is the preparation of detailed background reports on key issues. These media relations people also work to develop a range of anticipated questions and suitable answers for those who engage the media. This strategic approach affords some opportunity for molding the media coverage, on one hand. On the other hand, it means that a minister, or senior bureaucrat, may be thoroughly coached on appropriate responses that will drive the organization's preferred message in response to media questions.

In Canada, it is important to remember that millions of dollars are spent on media campaigns and public opinion polls on the way to an election and even during a party's term in office. The mass media have taken up a central position in our political landscape and cannot be avoided; nor can their impact be minimized. Likewise, in the controversial world of policing and public safety, police organizations (and the elected officials who have responsibility for these matters) must be prepared to make adequate investment in communications so that meaningful messages will be transmitted to the public.

It may be worthwhile noting that the nation's mass media outlets — excluding those organizations, such as the CBC, that are publicly owned and operated — are private corporations locked in rather intensive competition with one another for revenues. Therefore, the mass media have a strong ongoing interest in controversy and conflict, and both politicians and police executives need to be prepared for the intense scrutiny that accompanies the serious situations that arise every day within the policing environment.

However, it is imprudent for the police to take an adversarial attitude toward the media. There is much to be gained from making the media an ally in the process of communicating with the many constituencies served by the police service. Many Canadian police organizations have engaged the services of professional media relations personnel to manage and maintain an effective relationship with the news media and enhance their image with the public.

CONCLUSION

This chapter has examined the role of special interest groups in the area of policing and public safety. As pressure groups proliferate, they play an increasingly important role in driving the policy-making activities of political parties across Canada, particularly parties in power at both the federal and provincial levels. We have considered how several special interest groups have developed a strategic advocacy role for themselves and their members in the area of policing and public safety.

This chapter has also underlined the importance and prevalence of public opinion polling in Canadian politics and has listed several key market research organizations that undertake this work on behalf of political parties, special interest groups, and government agencies.

To illustrate the high stakes that are attached to policing and public safety issues, this chapter has considered several controversies that have arisen for politicians in Canada. There are many further examples of political controversies where the specific matters relating to policing have forced the resignation of a government minister or have plunged a bureaucratic department into scandal.

Finally, we have taken a brief look at the overwhelmingly important role of the mass media in the context of modern policing in Canada. The political process has been transformed by the power of mass media to penetrate our lives. In the area of policing and public safety, the public always holds a vital interest, and the mass media are well positioned to provide constant scrutiny of every detail pertaining to these matters.

REFERENCES

Belfall, Donald (1992). *Associations in Canada: future impact and influence.* Toronto : FARE/CSAE Foundation.

Biro, Fred et al. (2000). *Police executives under pressure: a study and discussion of the issues.* Ottawa : Canadian Association of Chiefs of Police (Police Futures Group study series; no. 3).

Black, E.R. (1982). *Politics and the news: the political functions of the mass media.* Toronto : Butterworths.

Brock, Kathy L. (ed.)(2002). *Improving connections between governments, nonprofit and voluntary organizations: public policy and the third sector.* Montreal : McGill-Queen's University Press.

Canada. Royal Commission on the Status of Women (1970). *Report.* Ottawa : Information Canada.

Coleman, W.D. and Grace Scogstad (1990). *Policy communities and public policy in Canada.* Toronto : Copp Clark Pitman.

Desbarats, Peter (1996). *Guide to Canadian news media. 2nd ed.* Toronto : Harcourt Brace.

Dyck, Rand (1993). *Canadian Politics: Critical approaches.* Scarborough, Ont. : Nelson Canada.

Ericson, R.V. et al. (1989). *Negotiating control: a study of new sources.* Toronto : University of Toronto Press.

Federation of Canadian Municipalities (2000). *A synopsis of a national symposium: towards a community response to organized crime.* Ottawa : Federation of Canadian Municipalities.

Goldstein, Jonah (1979). "Public interest groups and public policy: the case of the Consumers' Association of Canada." *Canadian Journal of Political Science,* March, pp. 137–155.

Griffin, David (2002). "Police association advocacy: a strategic priority." *The Canadian Police Association Express,* Issue 55, Spring, pp. 10–14.

Johnson, R. (1985). *Public opinion and public policy in Canada.* Toronto : University of Toronto Press.

Kwavnick, David (1972). *Organized labor and pressure politics.* Montreal : McGill-Queen's University Press.

Lachappelle, Guy (1992). *Polls and the media in Canadian elections: taking the pulse.* Toronto : Dundurn Press.

Malvern, Paul (1985). *Persuaders: lobbying, influence peddling and political corruption in Canada.* Toronto : Methuen.

Paltiel, K.Z. (1982). "The changing environment and role of special interest groups." *Canadian Public Administration,* Summer, pp. 198–210.

Pealow, James (2001). *Benchmarking and best practices for associations.* Toronto : Canadian Society for Association Executives.

Presthus, Robert (1973). *Elite accommodation in Canadian politics.* Toronto : Macmillan.

Pross, A. Paul (ed.) (1975). *Pressure group behaviour in Canadian politics.* Toronto : McGraw-Hill Ryerson.

———— (1990). "Pressure groups: talking chamelions." In Whittington, Michael and Glen Williams (eds.). *Canadian politics in the 1990s.* Toronto : Nelson Canada, pp. 285–290.

———— (1992). *Group politics and public policy. 2nd ed.* Toronto : Oxford University Press.

Reed, Wilson Edward (1999). *The politics of community policing: the case of Seattle.* New York : Garland Publishing, Inc.

Romanow, W.I. and W.C. Soderlund (1992). *Media Canada: an introductory analysis.* Toronto : Copp Clark Pitman.

Sawatsky, John (1987). *The insiders: power, money and secrets in Ottawa.* Toronto : University of Toronto Press.

Siegel, Arthur (1983). *Politics and the media in Canada.* Toronto : McGraw-Hill Ryerson.

Taras, David (1990). *The newsmakers: the media's influence on Canadian politics.* Scarborough, Ont. : Nelson Canada.

Thorburn, H.G. (1985). *Interest groups in the Canadian federal system.* Toronto : University of Toronto Press.

Trueman, Peter (1980). *Smoke and mirrors: the inside story of television news in Canada.* Toronto : McClelland & Stewart.

Vipond, Mary (1990). *The mass media in Canada.* Toronto : James Lorimer & Co.

White, Walter L., Ronald H. Wagenberg, and Ralph C. Nelson (1994). *Introduction to Canadian politics and government. 6th ed.* Toronto : Harcourt Brace Canada.

Wilson, J. (1980). "Media coverage of a Canadian election campaign: horserace journalism and the meta-campaign." *Journal of Canadian Studies,* Vol. 15, No. 4, pp. 56–68.

WEBLINKS

Special Interest Groups with Public Safety and Policing Concerns:

Alberta Civil Liberties Research Centre
www.aclrc.com

Amnesty International
www.amnesty.org

Assembly of First Nations
www.afn.ca

B.C. Civil Liberties Association
www.bccla.org

Canadian Association for Civilian Oversight of Law Enforcement
www.cacole.ca

Canadian Association of Police Boards
www.capb.ca

Canadian Bar Association
www.cba.org

Canadian Civil Liberties Association
www.ccla.org

Canadian Jewish Congress

www.cjc.ca

Canadian Labour Congress

www.clc-ctc.ca

Canadian Race Relations Foundation

www.crr.ca

Canadian Society of Association Executives

www.csae.com

Canadian Teachers' Federation

www.ctf-fce.ca

Federation of Canadian Municipalities

www.fcm.ca/english

John Howard Society

www.johnhoward.ca

Mothers Against Drunk Driving (MADD)

www.madd.ca

Canadian Polling Firms:

EKOS

www.ekos.com

Environics Research Group

www.environics.net

Goldfarb Consultants (Millward Brown)

www.millwardbrown.com

Hill & Knowlton Canada

www.hillandknowlton.ca

Ipsos-Reid

www.ipsos-reid.com/ca

QUESTIONS FOR CONSIDERATION
AND DISCUSSION

1. What restrictions, if any, should be placed on special interest, or pressure, groups in Canada?

_____

_____

_____

_____

_____

_____

_____

_____

2. How should groups of people organize themselves to put pressure on governments for change?

_____

_____

_____

_____

_____

_____

_____

_____

3. What guides the approach to media reporting on policing or public safety issues in newspapers, radio, and television?

_____

_____

_____

_____

_____

_____

_____

_____

_____

4. What difficulties may arise from too close a relationship between the members of the media and politicians, police executives, or senior bureaucrats?

_____

_____

_____

_____

_____

_____

_____

_____

_____

5. Public opinion polls are valuable to governing within a democracy. Do you agree, or disagree? Explain your answer.

_____

_____

_____

_____

_____

_____

_____

_____

RELATED ACTIVITIES

1. Complete the 3SC survey featured on the Environics Research Group website (see Weblinks). Once individuals have completed this survey, the instructor could facilitate in-class discussions on students' results. How could this type of survey instrument be adapted for use in a policing, public safety, or law enforcement setting?

2. Follow the coverage of a particular news item relating to policing or public safety in Canada in the various media outlets. How does the coverage of that item vary from one news medium to another? For example, is the coverage in the newspapers more detailed than the coverage provided on television? Are there regional differences in the coverage of a particular story? Can you detect any biases in the coverage?

3. Select one of the special interest groups listed in this chapter and examine its engagement in a particular issue relating to policing or public safety. For example, how has the Canadian Teachers' Federation dealt with the issue of youth violence? Or, what efforts has the Canadian Civil Liberties Association made to influence new anti-terrorist legislation?

4. Locate an example of a recent public opinion poll and consider the design of the questions, the approach to gathering data, the analysis of the results, and the presentation of the findings. Attempt to find public opinion polls that deal with policing or public safety issues. Alternatively, design a series of questions that might be used in a public opinion poll dealing with current policing and public safety concerns.

The Origins of
Public Administration

In chusing Persons for all Employments, they have more Regard to good Morals than to great Abilities: For, since Government is necessary to Mankind, they believe that the common Size of human Understandings, is fitted to some Station or other; and that Providence never intended to make the Management of publick Affairs a Mystery, to be comprehended only by a few Persons of sublime Genius, of which there seldom are three born in an Age: But, they suppose Truth, Justice, Temperance, and the like, to be in every Man's Power. . .

—Jonathan Swift, *Gulliver's Travels,* Part I, Chapter VI

Royal Newfoundland Constabulary.

Royal Newfoundland Constabulary officers on parade at Fort Townshend, with Imperial barracks in background, 1945.

Learning Objectives

1. Identify the theoretical basis for the study of public administration.
2. List several aspects pertaining to administrative responsibility in Canada.
3. Summarize several key problems relating to public management in Canada.

INTRODUCTION

In Chapter 1, "The Fundamentals of Political Science and Public Administration," we considered some of the essential elements of these two disciplines. This chapter will attempt to narrow the focus to the specifics of public administration as a distinct area of intellectual study. There will be some attention paid to the theoretical background for the investigation of matters pertaining to public administration. We will look at how public administration may be viewed as a concentrated, though not self-contained, perspective on a particular galaxy of issues.

Next, we will look at the concept of administrative responsibility and consider the relevance of this notion to policing and public safety in Canada.

Finally, this chapter will review some of the major problems, or challenges, currently facing those engaged in the realm of public management in Canada. This discussion will lay the groundwork for Chapter 9, "Public Policing — Critical Issues in Safety, Security, and Surveillance," where specific policing issues will be considered within the larger context of public management.

THEORETICAL BACKGROUND ON PUBLIC ADMINISTRATION

One of the great difficulties for public administration in Canada as a discipline has been its ongoing quest to firmly establish itself as a legitimate and coherent focus of independent study. Typically, in Canada, public administration has been subsumed, if not submerged, under the rubric of political science. Indeed, many of the existing university programs of public administration reside in their respective political science departments.

The essence of public administration may be seen as the study of systems, processes, and structures that are directed toward the public interest. Public administration concerns itself with the institutions, individuals, and instruments that work to achieve some form of common good. Obviously, it is quite simple to distinguish public administration from the management of private affairs. Clearly, there must be points of overlap between the public and private realms with regard to organizational, or administrative, principles. Still, it is worthwhile to think about what distinguishes the public interest from private interest. Indeed, this is the object of a recent work by the noted urban thinker Jane Jacobs (1992).

One distinguishing feature of public administration is the complex notion of accountability to citizens, or taxpayers. This can be compared with the private corporation's responsibility to shareholders and the guiding motivation of the desire for profit. Public administration is by and large guided by something we may call civic responsibility. There are standards of practice that tend to inform public administration, and these may help to differentiate this area of study and practice from other forms of management and administration.

It is also important to address the theoretical separation of the realm of public administration from the political sphere. The entire concept of an expert civil service, with its requirement for an array of competencies that are geared to the management of public affairs, helps to distance public administration from the spectre of partisan politics. The theory of public administration posits the existence of a value-free, objective, technically competent, perhaps even scientific approach to public affairs. Of course, much of the study of public administration examines the tension between this position and the realities of Canadian public management.

Several leading proponents of Canadian public administration have acknowledged that the theoretical underpinnings of this area of study are somewhat compromised (Kernaghan and Charih, 1996). However, it is not impossible to sketch some of the key areas of concentration for those engaged in the study and application of public administration in Canada. The following listing represents a series of fundamental topics that reliably inform this discipline:

- ethics in the public service;

- policy formulation, implementation, and evaluation;

- service delivery and compliance;

- financial management of public institutions;

- organizational theory;

- quantitative and qualitative research methods;

- political economy; and

- human resources management of public institutions.

While each of these topics may not, if taken individually, be exclusive to the study of public administration, when they are combined into an integrated curriculum of theoretical study and practical application, they do constitute the foundation of the art and science of public administration in Canada.

ADMINISTRATIVE RESPONSIBILITY

There are certain areas of administrative responsibility that are distinctive to public management. To understand the scope of concerns relevant to police organizations in Canada, it is worthwhile to consider the following topics:

Development and maintenance of a competent civil service. All public services, including police and public safety organizations, must be engaged in a process to ensure that their human resources possess the knowledge, skills, and abilities required to deliver the programs and services provided to the public.

Balance between discretion and accountability. It is important that those responsible for the administration of public affairs be conscious of their need to apply a degree of discretion to their functional role while simultaneously meeting the accountability demands of a responsible public service.

Proliferation of regulatory bodies and instruments. The role of government frequently requires the introduction of policy instruments that take the form of regulations. As civil society becomes more complex and demanding, the number of areas where government has a public interest increases dramatically. This is especially true in the area of policing and public safety, where people's security and well-being are often at stake. Accordingly, administrators must be conscious of the need for appropriate regulatory instruments and enforcement bodies, while also ensuring that the public is not overburdened with restrictive regulatory regimes.

Citizen engagement.   Democracy can only function properly when there is a genuine, ongoing, and informed engagement of the citizenry. This is one of the vital concerns of the philosophy of community, or problem-solving, policing. In all areas of public administration and management there is a requirement that public satisfaction be addressed and that the common good be a hallmark of administrative activities conducted on behalf of the public.

Administrative secrecy and freedom of information.   There are special constraints on public services to deliver their programs in a manner that will protect privacy, ensuring that administrative information is protected, while still ensuring freedom of information.

Ethical standards.   There is a pressing need for extensive ethical standards that may be applied across the public service, regardless of jurisdiction and level of engagement in that area of human endeavour. Distinct from the parameters of behaviour that might be applicable within the business world, and irrespective of the standards of conduct required of elected officials, public servants must be held to an extremely high code of behaviour that is consistent with their role as stewards of the public trust.

PROBLEMS OF PUBLIC MANAGEMENT

With the appearance of what has been referred to as the "new public management," a few questions have arisen. Practitioners and professors of public administration are both concerned with the improvement of public agencies. This requires the acquisition of sufficient knowledge to formulate and promote such improvements. Among the current issues facing public management are the following:

- anti-bureaucratic ethos among legislators;
- political neutrality;
- public service values of accountability, integrity, and equity;
- workability of the new public management paradigm;
- alternative service delivery models;
- gender issues;
- intergovernmental relations;
- adequacy of the policy capacity of governments;
- management of diversity issues;
- best practices versus worst practices;
- distinctions between policy formulation and policy execution;
- mechanisms for protecting taxpayers and senior officials;
- corruption-free, professional civil service.

As practitioners and academics of public management confront a new century, certain problems require special attention on both theoretical and practical fronts. To assist in this enterprise, there are several significant organizations that are pertinent to the careful study of public administration. These organizations have considerable value for police services, as well as public safety organizations, in reviewing and reflecting upon the challenges of public management. The following organizations warrant closer attention.

Canadian Centre for Management Development

Based in Ottawa, the Canadian Centre for Management Development (CCMD) is an organization that works to develop excellence in the management of the Canadian public sector. It has a mandate to provide educational and learning programs that will support current public service managers and work to develop future managers within the civil service. The CCMD accomplishes its mandate through courses, learning events, research, publications, and other activities designed to promote the development of leadership.

The CCMD maintains an incredibly extensive website that includes features on leadership development, public sector management, and "e-government" processes. This website includes a broad range of reports and publications produced under the auspices of the CCMD, as well as speeches delivered by the organization's president, which frequently address pivotal issues relevant to public servants across Canada.

Carleton University

The School of Public Policy and Administration at Carleton University is extremely active in the field of study that pertains to public management. It offers undergraduate and graduate programs that are tailored to the needs of students who wish to pursue careers in the public service, policy development, and other areas related to public affairs.

The *Carleton Journal of Public Administration* is an online publication that features the academic work of graduate students in public administration and related fields of policy study. It is a valuable forum for the exchange of insight in relevant areas of study and promotes dialogue within a community that includes academics, researchers, and policy analysts.

The Arthur Kroeger College of Public Affairs is also located within Carleton University and focuses its efforts on the delivery of an interdisciplinary program that adheres to a core curriculum designed to develop an understanding of public affairs and policy management.

Conference Board of Canada

The Conference Board of Canada is a national organization that seeks to build leadership capacity for improving Canada. This goal is accomplished through the formulation and dissemination of insights dealing with economic trends, public policy, and organizational performance. All of these "knowledge areas" have direct application to policing and public safety organizations and are important to the growth of sound public management in Canada.

The Conference Board of Canada was created in 1954 and is an independent, non-profit applied research organization that sponsors a wide range of events in support of learning and network building. The Conference Board of Canada also publishes a wealth of research products and provides customized research services to its clients in both the public and private sectors.

The Conference Board of Canada sponsors the Canadian Forum for Learning and Development. This forum examines sound policies and practices in corporate training, education, and development. The forum is intended to offer insights and guidance that will enhance the capacity of member organizations to adopt effective management and leadership practices. It is worthwhile to note that the Royal Canadian Mounted Police is among the many corporate member organizations affiliated with this forum.

The Conference Board of Canada also supports the work of the Public Sector Executives Network. This body is designed to provide opportunities for senior executives in government departments and agencies to share insights and learning that may contribute to strengthening their organizational effectiveness. There is an examination of best practices that may be translated into various provincial settings, and members consider innovative approaches to public sector reform. Also, the Conference Board of Canada has established a Government Performance and Accountability Network that is designed to provide guidance and assistance in the definition and implementation of accountability and performance measurement systems within government. Again, the Royal Canadian Mounted Police has an organizational membership in this network, in line with its efforts to generate fresh thinking and practical approaches to performance and accountability challenges.

Institute of Public Administration of Canada

The Institute of Public Administration of Canada (IPAC) is Canada's foremost organization dedicated to the promotion of excellence in public management and public service. Its membership includes academics, practitioners, students, and various service partners representing all levels of government and all regions of the country.

IPAC undertakes a wide range of activities designed to promote and celebrate the highest standards of professional practice and service in the realm of public administration. It offers many benefits to its individual and organizational members, including the following:

- publication of *Canadian Public Administration,* the official journal of IPAC (this professional and scholarly journal includes articles of interest and importance to practitioners and academics in the area of public management);

- publication of substantive monographs and series of essays that examine issues of relevance to members of IPAC and those interested in public administration;

- publication of *Public Sector Management,* a magazine that deals with current matters of concern to public managers and administrators;

- development of an annual conference built around a central theme, with speakers and sessions, designed to enhance understanding of important public administration issues;

- workshops, seminars, and other learning events that promote the expansion of knowledge in the area of public management;

- conducting research on critical issues concerning the public administration community; and

- maintaining an up-to-date website to keep members informed, engaged, and enlightened.

On a regular basis, IPAC conducts a survey of deputy ministers, and others, across the country in order to determine the key public management issues they expect to face in their working lives. In the year 2000, the following "major themes" emerged from the surveys completed:

- renewing the human dimension of public management, including organizational performance and health;

- redesigning and enhancing service delivery from a citizen-centred perspective;

- reframing accountability arrangements and expanding the application of performance measurement; and

- rethinking and reshaping government's role, size, and policy capacity within fiscal targets and managing organizational changes that result.

There is a pressing need for greater attention to organizational learning, including the management of knowledge. This entails building knowledge systems that place an emphasis on public servants as "knowledge workers" and provides them with the wherewithal to generate problem-solving approaches to the challenges of public management. This will ensure high performance within the constantly changing work environment.

The Fraser Institute

The Fraser Institute is an independent public policy organization that concentrates on the role of competitive markets in creating economic and social well-being for Canadians. Its research focus turns toward educational, environmental, regulatory, fiscal, and related areas. Frequently The Fraser Institute will examine governance issues that are pertinent to policing and public safety organizations across Canada. The institute has examined areas including Canada's institutions of governance, charters, federalism, democratic reform, land use, and Aboriginal policy. Many of these research projects have significance for the strategic planning that takes place in police services and public safety organizations. The *Fraser Forum,* the institute's monthly review of public policy in Canada, is valuable for students of public administration and often addresses topics relating to the criminal justice system in Canada.

CONCLUSION

This chapter has briefly sketched the theoretical background pertaining to public administration in Canada. It has considered some of the linkages between the academic study of public management and other academic disciplines such as political science and management studies. We have considered the major areas of administrative responsibility that engage public servants in Canada and which continue to be relevant to policing and public safety organizations.

Finally, we have taken an opportunity to examine some of the key problems currently confronting public administration in Canada today. This has included the agenda for research and study that has been articulated by senior public executives and others who have an academic interest in this field of endeavour. We have briefly presented some of the significant organizations and institutions that have the enhancement of public administration and management as a cornerstone of their mandates.

The task of the public administrator is both challenging and daunting. In a time when citizens, as taxpayers and consumers, are demanding more and better services from their governments, it is essential that public servants be equipped to understand and address those demands. Ironically, this is also a time when there are equally compelling pressures on the public sector to refine the focus of the services they deliver and to seek out alternative delivery methods that may, or may not, include the private and volunteer sectors. These challenges can be detected throughout the public service and at all levels of government. They have a profound impact on policing and public safety organizations, as we will examine in more detail in Chapter 9 ("Public Policing — Critical Issues in Safety, Security, and Surveillance").

REFERENCES

Aucoin, Peter (1996). *The new public management: Canada in comparative perspective.* Montreal : Institute for Research on Public Policy.

Barberis, Peter (1998). "The new public management and a new accountability." *Public Administration,* Vol. 76 (Autumn), pp. 451–470.

Bellamy, Christine and John A. Taylor (1998). *Governing in the information age.* London : Open University.

Borins, Sandford (1995). "The new public management is here to stay." *Canadian Public Administration,* Vol. 38, No. 1 (Spring), pp. 122–132.

Bozeman, Barry (1987). *All organizations are public: bridging public and private organizational theories.* New York : Jossey-Bass.

Clifford, John and Kernaghan Webb (1986). *Policy implementation, compliance and administrative law.* Ottawa : Law Reform Commission of Canada.

Cohen, Steven and William Eimicke (2002). *The effective public manager. 3rd ed.* New York : Jossey-Bass.

Condrey, Stephen E. (ed.) (1998). *Handbook of human resource management in government.* San Francisco : Jossey-Bass.

Doern, G. Bruce (1994). *The road to better public services: progress and constraints in five federal agencies.* Montreal : Institute for Research on Public Policy.

Ford, Robin and David Zusman (eds.) (1997). *Alternative service delivery: sharing governance in Canada.* Toronto : Institute of Public Administration of Canada.

Gruber, Judith E. (1988). *Controlling bureaucracies.* Berkeley : University of California Press.

Hall, Michael H. and Paul B. Reed (1998). "Shifting the burden: how much can government download to the non-profit sector?" *Canadian Public Administration*, Vol. 41, No. 1 (Spring), pp. 1–20.

Hood, Christopher (1983). *The tools of government.* London : Macmillan.

——— (1996). "Control over bureaucracy: cultural theory and institutional variety." *Journal of Public Policy,* Vol. 15, No. 3, pp. 207–230.

Jacobs, Jane (1992). *Systems of survival: a dialogue on the moral foundations of commerce and politics.* New York : Random House.

Keehley, Patricia et al. (1996). *Benchmarking for best practices in the public sector: achieving performance breakthroughs in federal, state, and local agencies.* New York : Jossey-Bass.

Kernaghan, Kenneth (1968). "An overview of public administration in Canada today." *Canadian Public Administration,* Vol. 11, No. 3 (Fall), pp. 291–308.

——— (1977). *Public administration in Canada: selected readings. 3rd ed.* Toronto : Methuen.

——— (1982). "Canadian public administration: progress prospects." *Canadian Public Administration,* Vol. 25, No. 4 (Winter), pp. 232–256.

————— (1993). "Reshaping government: the post-bureaucratic paradigm." *Canadian Public Administration,* Vol. 36, No. 4 (Winter), pp. 636–645.

Kernaghan, Kenneth and Mohamed Charih (1996). *Report of the national conference on research in public administration: an agenda for the year 2000.* Ottawa : Canadian Association of Programs in Public Administration.

Kernaghan, Kenneth, Brian Marson, and Sandford Borins (2000). *The new public organization.* Toronto : Institute of Public Administration of Canada.

Light, Paul Charles (1999). *The new public service.* Washington, D.C. : Brookings Institute.

Linden, Russell M. (1994). *Seamless government: a practical guide to re-engineering in the public sector.* San Francisco : Jossey-Bass.

Lynn, Laurence E. (2001). "The myth of the bureaucratic paradigm: what traditional public administration really stood for." *Public Administration Review,* Vol. 61, No. 2 (March/April) pp. 186–203.

March, James G. and Johan P. Olson (1989). *Rediscovering institutions: the organizational basis of political life.* New York : Free Press.

Perrow, Charles (1986). *Complex organizations: a critical essay. 3rd ed.* New York : McGraw-Hill.

Perry, James L. (ed.) (1996). *Handbook of public administration. 2nd ed.* San Francisco : Jossey-Bass.

Peters, Guy (1989). *The politics of bureaucracy. 3rd ed.* New York : Longmans.

Rainey, Hal G. (1996). *Understanding and managing public organizations. 2nd ed.* San Francisco : Jossey-Bass.

Savoie, Donald J. (1995). "What is wrong with the new public management?" *Canadian Public Administration,* Vol. 38, No. 1 (Spring), pp. 112–121.

Seidle, Leslie (1996). *Rethinking the delivery of public services to citizens.* Montreal : Institute for Research on Public Policy.

Sossin, Lorne (1993). "The politics of discretion: toward a critical theory of public administration." *Canadian Public Administration,* Vol. 36, No. 3 (Fall), pp. 364–391.

Sutherland, S.L. (1991). "The Al-Mashat affair: administrative responsibility in parliamentary institutions." *Canadian Public Administration,* Vol. 34, No. 4, pp. 573–603.

Wholey, Joseph S. and Kathryn E. Newcomer (1989). *Improving government performance: evaluation strategies for strengthening public agencies and programs.* San Francisco : Jossey-Bass.

Wilson, James Q. (1989). *Bureaucracy.* New York : Basic Books.

Yates, Douglas (1991). *The politics of management: exploring the inner workings of public and private organizations.* San Francisco : Jossey-Bass.

WEBLINKS

Canadian Centre for Management Development

www.ccmd-ccg.gc.ca

Carleton Journal of Public Administration

www.carleton.ca/mpa/cjpa

Carleton University, Arthur Kroeger College of Public Affairs

www.carleton.ca/akcollege

Carleton University, School of Public Policy and Administration

www.carleton.ca/spa

Conference Board of Canada

www.conferenceboard.ca

Institute of Public Administration of Canada

www.ipaciapc.ca

The Fraser Institute

www.fraserinstitute.ca

QUESTIONS FOR CONSIDERATION AND DISCUSSION

1. What are the key distinguishing features of the study of public administration in Canada?

   _____

   _____

   _____

   _____

   _____

   _____

2. What are the main characteristics of the "new public management"?

   _____

   _____

   _____

   _____

   _____

   _____

3. Where would the issues relevant to public administration have the most impact upon Canadian police organizations?

   _____

   _____

   _____

   _____

   _____

   _____

4. How would the elements of administrative responsibility outlined above influence police and public safety organizations in Canada?

_____

_____

_____

_____

_____

_____

_____

5. What problems identified for public administration would be of significant concern for Canadian police organizations? What are some remedies that might be applied to these problems?

_____

_____

_____

_____

_____

_____

_____

_____

6. How would an in-depth study of public administration benefit the leadership or management of modern police organizations?

_____

_____

_____

_____

_____

_____

_____

_____

_____

RELATED ACTIVITIES

1. Review current and back issues of *Canadian Public Administration*. Discuss the various topics that have been addressed in this official journal of the Institute of Public Administration of Canada. Note the blend of both theoretical and practice-oriented essays and articles.

2. Locate specific articles in *Canadian Public Administration* relating to policing and public safety issues.

3. Review course outlines developed by schools of public administration in Canada. Consider the focus of these courses and discuss the overlap with programs in political science, management studies, or other areas.

4. Review the Institute of Public Administration of Canada's Award for Innovative Management (2002), and consider some of the finalist projects that have been submitted for this award. Pay particular attention to project submissions that deal with policing and public safety, e.g., Alberta Solicitor General, Youth Justice Community Project, and Alberta Transportation Inspection Services Branch CALEA Accreditation Project.

chapter seven

Canadian Policy-Making in Action

No one takes to public affairs on official instruction or compulsion. When a man is sufficiently persuaded of his ability to come forward, we act as good sense and humane feeling dictate, we accept him with friendly good will, we elect him to public position and trust him with the conduct of affairs. A man who succeeds earns greater kudos and has the pull over others for that reason. If he fails, will he offer pretexts and excuses? Justice precludes this.

—Demosthenes, *Demosthenes and Aeschines* (translated by A.N.W. Saunders)

Royal Newfoundland Constabulary.

Royal Newfoundland Constabulary officer directing traffic at the corner of Prescott and Duckworth streets, St. John's, circa 1940.

Learning Objectives

1. Understand the fundamentals of policy analysis and development in Canada.
2. Identify the 10 components of the public policy cycle.
3. List the main policy instruments available to decision-makers.
4. Identify the major organizations and institutions that support and promote the development of public policy in Canada.
5. List key publications that focus on public policy issues.

INTRODUCTION

This chapter will examine some of the fundamentals of policy analysis and development. It will present an overview of the public policy cycle, dividing the entire process into 10 main components and explaining each component as it relates to the realm of policing and public safety in Canada.

This chapter will also look at some of the organizations and individuals of the policy community that deal with topics of relevance for this area of study. It will focus exclusively on police organizations in Canada that have developed some capacity to undertake this task. However, it is important to be aware that policy-making in general is a much broader topic that incorporates many public, private, and voluntary organizations.

We will also consider the range of policy instruments available to decision-makers for realizing their public policy goals.

Policy coordination will be looked at from the perspective of the government's role in making sure that all loose ends are tied together. Finally, we will look at some of the organizations and institutions that take some responsibility for the whole process of policy review and evaluation in Canada.

FUNDAMENTALS OF POLICY ANALYSIS AND DEVELOPMENT

Public policy is an important field of study because it represents the framework within which governments put their ideas into action. To achieve effective management in public affairs, those in positions of power must have some policy perspective that will inform how they operate and what will, ultimately, constitute success. Therefore, the process of developing, implementing, and evaluating public policy is of vital interest to academics, bureaucrats, Cabinet ministers, and citizens alike.

There is a vast array of factors that may contribute to the development of public policy in Canada. Brooks (1993) explains the intricacies of the policy-making process as follows:

> Policy-making is filtered through a complex institutional framework. The differences between the institutional arrangements characterizing capitalist democracies are quite significant. Those differences affect how political conflict is expressed, what strategies individuals and groups will employ in attempting to influence policy, and the weight that policy-makers ascribe to particular social and economic interests. *(p. 79)*

Government officials at all levels will frequently engage policy analysts to undertake empirical or theoretical research that will assist them in the formulation of policy options on any given topic. Frequently, these officials will turn to their own bureaucracy for this purpose; however, increasingly, there is a trend toward engaging the services of outside consultants and public policy think-tanks to review a particular issue and provide policy recommendations for government consideration. This trend is part of a movement away from in-house policy work to the engagement of professional researchers for the purposes of detailed policy development. It is also important to take note of the significant role played by various royal commissions, task forces, and commissions of inquiry. These special-purpose undertakings are established to intensively examine a particular public policy issue and to offer a comprehensive range of conclusions and recommendations that government may take under advisement when developing its official response to that issue.

As the complexity of life in a modern, liberal democratic society like Canada increases, it is possible to witness the growing influence of various disciplines in the process of policy analysis. When one considers the enormous range of issues and enterprises regulated by the various levels of government, it is not surprising that decision-makers must turn to the following disciplines for substantial guidance and direction:

- political science

- economics

- psychology

- sociology

- law

- criminology

- any other discipline that may be relevant to the policy issue being examined

Within the vast realm of concerns dealt with by government departments, several fundamental dimensions of public policy may be differentiated. They include the following basic capacities of government:

- *Coercive.*   This includes the capacity of governments to apply compelling political authority, as with the use of sanctions.

- *Distributive.*   This includes the capacity of government to allocate goods, services, rights, and privileges both directly and indirectly, as well as to allocate offices and positions of responsibility within the political infrastructure.

- *Systemic.*   This includes the capacity of government to exercise its authority across the entire spectrum of public affairs.

Furthermore, in the field of Canadian public policy, it is possible to identify 10 fundamental phases through which an identified policy issue will pass in the normal course of events. Figure 7.1 presents the public policy cycle as follows:

To better understand the 10 phases of this cycle, we will look at each element and consider its key characteristics.

- *Policy issue identification.*   This is where a particular public policy matter is identified as being of some concern or importance to those who have responsibility for some area of policy development. For example, once the value of DNA testing was established for law enforcement and policing purposes, it became important to develop policy for the application of this new technology across Canada. It was also essential that this policy be addressed in a manner that would allow it to be implemented with a high degree of consistency in all Canadian police jurisdictions. Accordingly, the policy issue of DNA testing became a matter for the federal Solicitor General's office and would involve police agencies at the federal, provincial, and municipal levels.

- *Policy issue acceptance.*   Once a public policy issue has been identified, it then becomes necessary for that issue to be accepted as a legitimate area of concern. If a particular policy issue has been identified but not accepted by decision-makers in

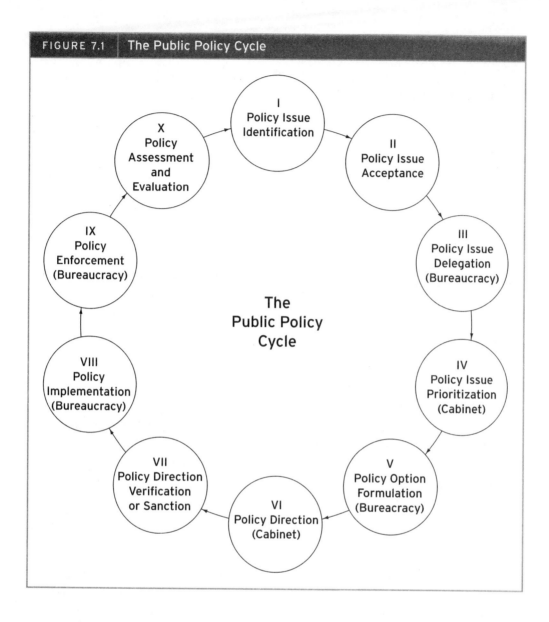

FIGURE 7.1   The Public Policy Cycle

The Public Policy Cycle

I Policy Issue Identification

II Policy Issue Acceptance

III Policy Issue Delegation (Bureaucracy)

IV Policy Issue Prioritization (Cabinet)

V Policy Option Formulation (Bureacracy)

VI Policy Direction (Cabinet)

VII Policy Direction Verification or Sanction

VIII Policy Implementation (Bureaucracy)

IX Policy Enforcement (Bureaucracy)

X Policy Assessment and Evaluation

that policy field, little work will be devoted to that issue. For example, the issue of capital punishment may be viewed by certain pressure groups as important from their perspective; however, if government decision-makers do not accept this as a legitimate issue for further attention, very little decisive action will be taken on this matter.

- *Policy issue delegation.*   This is where some level of government has identified and accepted a policy issue as legitimate and seeks to have further work done on the matter for possible action. Typically, the policy issue will be delegated to the bureaucracy within the department that has responsibility for this particular area of concern. For example, if the immediate matter concerns the use of pepper spray, or some other use of force option, there will be units or individuals within the federal Solicitor General's department with the skills, knowledge, and abilities to review this issue and provide guidance and information for the decision-makers. There may be, in such instances, a requirement for several areas within the bureaucracy to cooperate on the review and study of a particular policy issue to ensure that it is considered in a comprehensive manner. Alternatively, the government department responsible for this policy issue may engage the research services of an outside agency or institution to conduct a policy review on the department's behalf and provide that department with a full report, including policy recommendations for review.

- *Policy issue prioritization.* This is where the decision-makers determine the importance of a particular policy issue with respect to the government's overall agenda. Typically, there are so many serious matters that need to be addressed by the various levels of government, it is imperative that there be some degree of prioritization so that the work of government is done with some degree of efficiency and effectiveness. At this stage, the decision-makers (for example, the federal Cabinet) will review the various policy initiatives that are being presented to them and will establish some form of rank ordering that will permit the business of government to be conducted in some form of priority. This amounts to something like a threat assessment for government. The bureaucracy will often be required to provide decision-makers with an outline of an identified policy issue that includes the risks associated with any delay in acting upon the matter. For example, the results of the attacks that took place on September 11, 2001, caused the entire focus of the federal government to shift significantly to matters of public safety and the control of possible terrorist threats. This policy issue demanded the full attention of the federal Minister of Justice and the Solicitor General of Canada. However, it also rose to the top of the priority list for the entire federal Cabinet as government's undivided focus was placed on the effort to reassure the Canadian public that everything possible was being done to protect and preserve the integrity of our nation.

- *Policy option formulation.* This is where the bureaucracy provides decision-makers with a range of possible options with respect to a particular policy matter. For example, in the area of high-speed pursuits by police, the subject matter experts within the appropriate department(s) would provide a detailed outline of the issues involved and a range of possible policy options relevant to this matter. Normally, the discussion of each option would include an assessment of the benefits and possible adverse consequences of the option under consideration. Frequently, a full analysis includes the consequences of maintaining the status quo. It is during this phase that the bureaucracy will often include some consideration of the range of policy instruments that decision-makers may choose to invoke for addressing the particular policy matter. For example, in the case of high-speed pursuits by the police, the range of policy instruments could include legislation, regulation, fines, public information, etc.

- *Policy direction.* This is where decision-makers have an opportunity to select a particular option that has been presented by the bureaucracy. Once a policy option has been selected, the decision-makers will then direct that steps be taken to put that option into effect. This will typically require that the bureaucracy conduct further work to finalize the full details of the selected policy option. Once a specific policy direction has been endorsed by the decision-makers, it is important for the bureaucracy to ensure that all pertinent aspects of the policy initiative have been completed. This would include any preliminary work (for example, legislative drafting, policy statements, communications plans) that has been assigned to put the selected option into effect.

- *Policy direction verification or sanction.*   This is where the decision-makers formally approve of the selected policy direction, including the specific policy instrument(s) that will realize the substance of this policy decision. For example, when the Ontario Government determined that it would establish a sex offender registry, it was necessary to pass enabling legislation that would formally put into place the elements required for this purpose. Accordingly, *Christopher's Law (Sex Offender Registry)* was passed by the provincial legislature in 2000.

- *Policy implementation.*   This is where the bureaucracy puts the policy direction into effect and ensures that the elements approved by the decision-makers are established and functioning properly. When the policy instrument selected involves new legislation, the bureaucracy will normally assign individuals or units to the task of monitoring and facilitating the implementation of this direction. For example, when it was determined that the Canadian public needed better protection from sexual offenders, the federal Solicitor General announced that the Royal Canadian Mounted Police (RCMP) would be implementing a distinct sex offender category within the Canadian Police Information Centre (CPIC) system by the end of 2002. This new database, in conjunction with the existing sex offender registry, seeks to provide better protection for the public with regard to this category of offender. The information posted by police forces across Canada will be made available on the CPIC system. Any record in this new category will automatically generate a message to the provincial Violent Crime Linkage Analysis System (ViCLAS) for any necessary police follow-up.

- *Policy enforcement.*   This is where the decision-makers delegate responsibility to elements of the bureaucracy to provide whatever is required to ensure that the provisions of any policy instrument are met by those who are affected by the policy in question. For example, in the case of new firearms registration legislation in Canada, it became the responsibility of the federal Department of Justice to ensure that these provisions were enforced across Canada. As a result, the Canadian Firearms Centre was established to enforce the provisions of the *Firearms Act* throughout the country.

- *Policy assessment and evaluation.*   This element includes all those techniques and approaches that take into account the effectiveness of any particular policy direction. This phase essentially closes the loop on the public policy cycle. Its main purpose is to determine where there might exist gaps or flaws in the implementation of this policy direction. This phase of the cycle is very often minimized, or even ignored, in the public policy process. However, it represents a critical opportunity for decision-makers to determine whether the policy initiative has indeed met its goals and objectives. A competent evaluation and assessment will help to determine what sorts of adjustments might be required to more fully address the policy concerns that originally required this intervention. This phase may also include the judicial interpretation of policy directions approved by the decision-making bodies (e.g., Parliament).

Another model of policy analysis developed by Bardach (2000) presents an eightfold path including the following steps:

1. defining the policy problem;
2. assembling evidence relevant to the problem;

3. constructing various alternative solutions;

4. selecting criteria to evaluate the solution;

5. projecting outcomes expected from the solution;

6. confronting trade-offs associated with the solution;

7. deciding on a policy solution; and

8. telling the story of your solution.

POLICY COMMUNITIES IN THE REALM OF POLICING AND PUBLIC SAFETY

In Chapter 5 ("Policing in a Political Context") we looked at the vast network of organizations and individuals that may constitute the policy community in various areas of public interest. In this chapter we will concentrate on those components that are more closely aligned with the public service dimension of policing and public safety. Providing this more focused attention to the public organizations that pursue policy initiatives in this area does not imply that the other actors within the policy community are less important; however, it provides a solid foundation for understanding the public service axis in this area of activity.

At all levels of government — federal, provincial, territorial, regional, and municipal — some department, branch, or unit is tasked with responsibility for policy issues that pertain to policing and/or public safety. With so many resources, both human and financial, dedicated to the various aspects of these areas of activity, it is easy to understand the requirement for careful policy planning, implementation, and evaluation. Much of the Canadian criminal justice system rests upon a foundation of federal legislation. Therefore, it is also quite clear that a great deal of policy coordination is required within several federal government departments, between various levels of government (e.g., federal and provincial, inter-provincial, and provincial and municipal), and among the whole range of public safety and police service providers across Canada.

At the functional level, most police organizations, regardless of their size, have some capacity for undertaking the critical policy work that is necessarily required of the department. Because of the inherent complexity of these services, as well as the constantly changing nature of the legislative and regulatory framework within which these police departments operate, ongoing policy planning is a fact of life for modern Canadian police organizations.

To highlight the central importance of policy planning within the fabric of Canadian police services, several examples are provided below. Each of these departments has built a capacity to undertake the detailed policy work required to ensure that their service is consistent with the laws of Canada and with the legislative and regulatory requirements that prevail within their own jurisdiction. While each police service may situate this policy function within its organizational structure in a different way, all police services must involve themselves in activities that partake of some version of the public policy cycle outlined above in this chapter.

Royal Canadian Mounted Police

The RCMP, through its Commissioner, is under the overall direction of the Solicitor General of Canada. The RCMP is deployed across the country and serves as the federal police service in every province and territory. Additionally, the RCMP contracts its policing services on a provincial and municipal basis, as required. In order to coordinate such a broad base of operations, the RCMP is organized on the basis of a functional and regional management system that has the following structure:

- Atlantic Region (Halifax)
- Central Region (Ottawa)
- North West Region (Regina)
- Pacific Region (Vancouver)
- Operations
- National Police Services and Infrastructure
- Corporate Management and Comptrollership
- Strategic Direction

For the purposes of this chapter, the Strategic Direction portfolio is of particular interest. Headed by a Deputy Commissioner, this area includes the following divisions:

- Strategic Policy and Planning
- Public Affairs and Information
- National Communications
- Organizational Strategy

This component within the organizational structure of the RCMP is critical because it holds responsibility for the provision of advice and support to the senior executive in establishing strategic direction for the entire organization. Strategic direction includes ensuring that the goals and objectives of the RCMP are appropriately aligned with priorities across the entire federal government. The RCMP must understand and adapt to take into account environmental trends as well as the needs of its clients and strategic partners.

Accordingly, the functional area identified as Organizational Strategy is tasked with conducting environmental scanning on behalf of the RCMP that will capture data on economic, social, political, legal, technological, and other trends that may affect the organization. Based on such scanning, the area undertakes policy development and research that allows the RCMP to appropriately respond to key trends.

As an organization, the RCMP has adopted a strategic framework that facilitates the policy planning process. Consistent with its overall operational goal of "Safe Homes, Safe Communities," the RCMP has articulated its organizational strategy in the following categories:

- strategic priorities, including organized crime, youth, alternative justice, and international police services (peacekeeping);

- strategic objectives, including prevention and education, intelligence, investigation, enforcement, and protection; and

- management strategies, including intelligence, values, bridge-building, and accountability.

We have spent some considerable time examining the structure of the policy planning process within the RCMP. The elements of this process are pertinent to virtually any police organization, however, and therefore this section should be seen as broadly applicable. The need to carefully orchestrate the policy development process is illustrated in the structural arrangements that have been made by the RCMP to coordinate these important functions.

Ontario Provincial Police

As the provincial police service for Canada's most populous province, the Ontario Provincial Police (OPP) is structured in ways that are quite similar to the RCMP. This police service is under the authority of a Commissioner, who in turn reports to the Deputy Minister of Public Safety and Security.

Within the office of the Commissioner are the following divisions:

- Strategic Services

- Corporate Services

- Field and Traffic Services

- Investigation/Organized Crime Section

- Drug Enforcement Section

A key component of the Strategic Services division is the Operational Planning and Research Bureau. In this bureau resides the core of the OPP's capacity to undertake policy development projects. In attempting to pursue a continuous improvement model, the OPP uses this bureau to anticipate environmental trends and to undertake research that will permit senior management within the OPP to select appropriate organizational directions.

The Operational Planning and Research Bureau contains the following sections:

- Statistical Analysis and Research

- Operational Policy

- Major Projects and Community Policing

- Planning Section

Each of the sections identified above is designed to provide a strong policy planning capacity for this organization as it meets the ongoing administrative and operational challenges of policing in Ontario.

Ottawa Police Service

Following an amalgamation project in 1995, the Ottawa Police Service has grown in size and jurisdiction to provide policing services to the City of Ottawa as well as to the former jurisdictions of Gloucester and Nepean. As a significant municipal police service in the nation's capital, it is required to function in a complex environment with many challenges and crises to manage. Figure 7.2 shows the organizational structure of the Ottawa Police Service.

Within the Executive Services division of the Ottawa Police Service, which reports directly to the Chief of Police, we find the following relevant areas of activity:

- *Corporate Planning.* The responsibilities of this area include organizational development, resource analysis, problem-solving coordination, crime analysis, and crime prevention through environmental design.

- *Legal Services.* The responsibilities of this area include professional standards and policy development.

Peel Regional Police Service

This police service located in the suburbs immediately west of Toronto was established in January 1974 to coincide with the creation of the Regional Municipality of Peel. The new police organization incorporated the former police services of Brampton, Chinguacousy, Mississauga, Port Credit, and Streetsville. It is currently the second largest municipal police service in Ontario and serves a population base of more than 935,000. To ensure effective policy planning and development, the Peel Regional Police Service has several elements within its organizational structure that provide this function. Figure 7.3 represents the organizational structure of this police service.

The Chief Administrative Officer is responsible for Corporate Services. The Information Services department is part of Corporate Services and has the following area:

- Organizational Development, which undertakes corporate analysis, corporate registry, and organizational memory.

The Deputy Chief of Police for Operations is responsible for the following area:

- Operational Planning and Resources, which undertakes activities relating to research and development, emergency planning, coordination of crime analysis, and special projects.

POLICY INSTRUMENTS

Policy instruments are the specific techniques, devices, or approaches selected to deal with a particular public policy issue. These instruments permit government officials to operate in a variety of contexts, including the private sector, to influence, guide, and regulate a wide range of human activities.

Among the policy instruments that are available to government decision-makers are any, or all, of the following:

- maintaining the status quo

- privatization of the conflict

- exhortation and information

- tax expenditures

- public expenditures

- regulation

- taxation

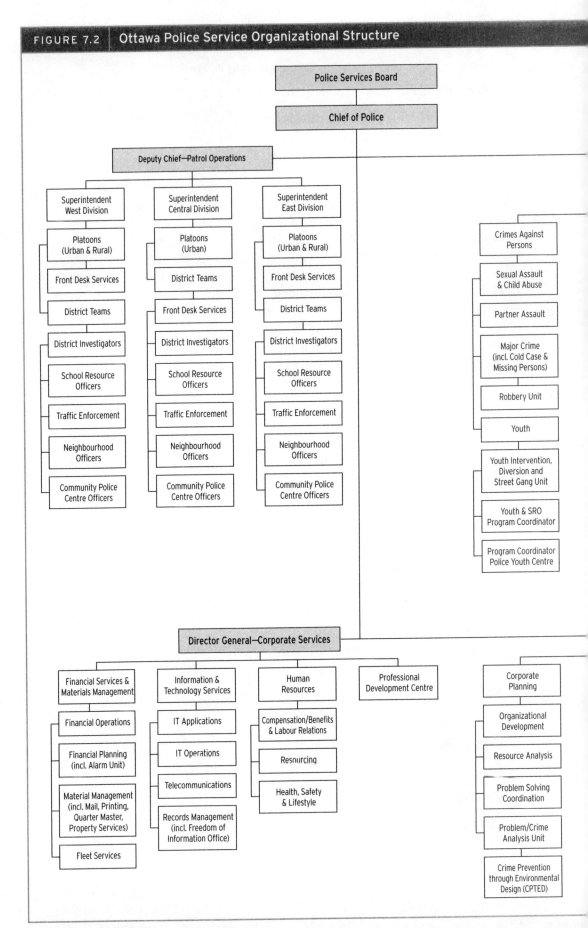

FIGURE 7.2 | Ottawa Police Service Organizational Structure

Source: Ottawa Police Service. Used with permission.

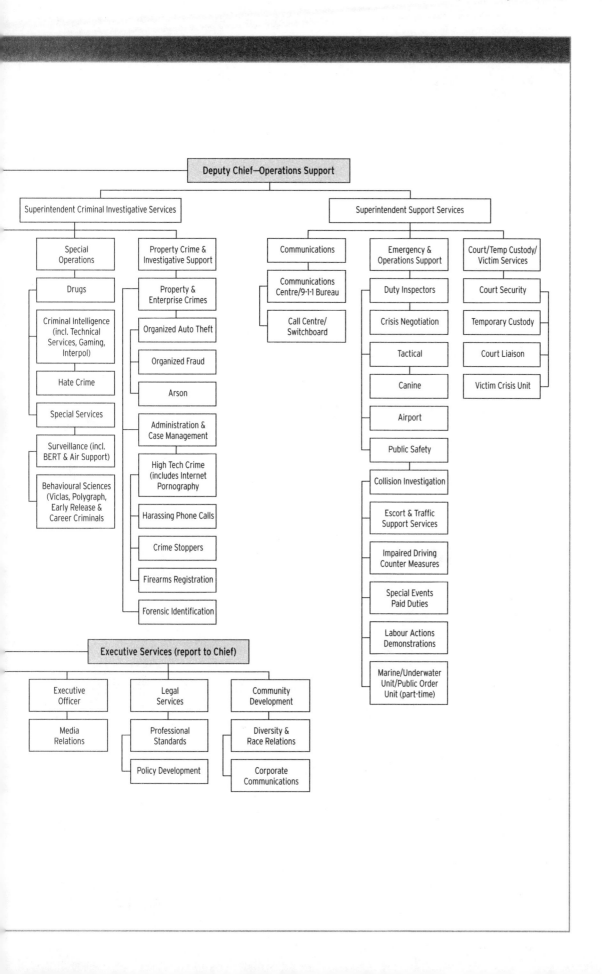

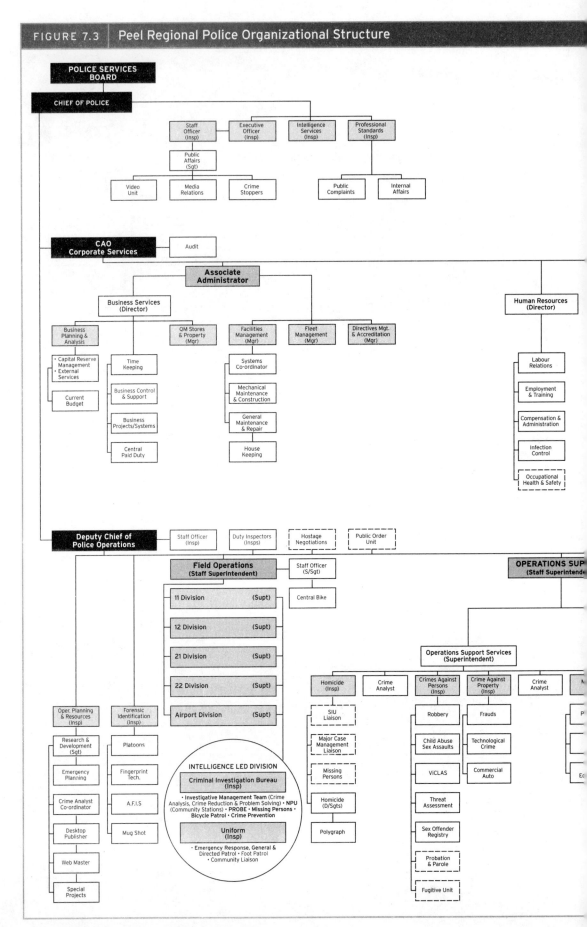

FIGURE 7.3 | Peel Regional Police Organizational Structure

Source: Peel Regional Police Service. Used with permission.

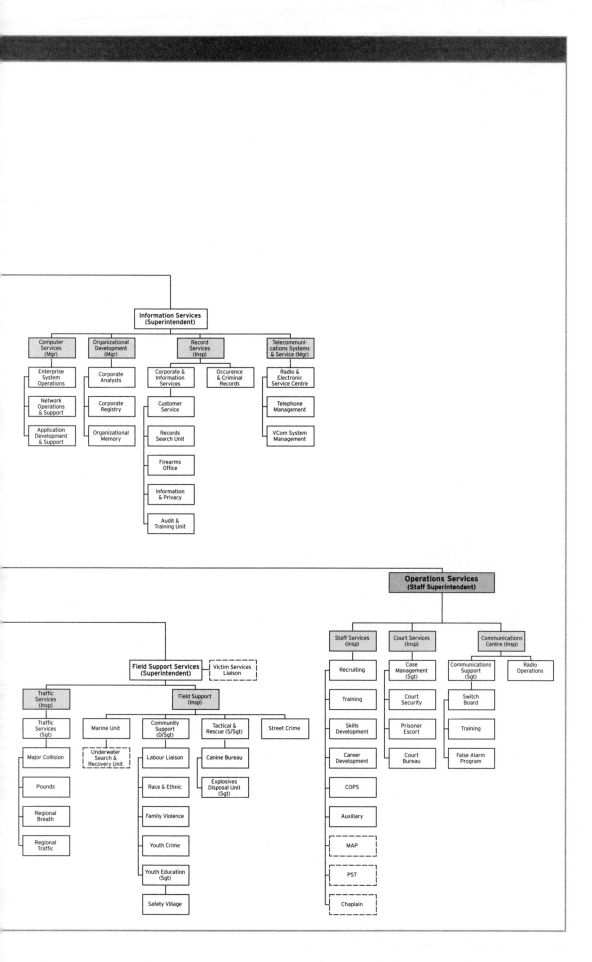

- public ownership
- state of emergency

The area of regulation, as a policy instrument, is particularly important in the realm of policing and public safety. As Dyck (1993) points out:

> Regulation is the hidden half of government, so to speak, and the regulatory process and the behaviour of regulatory agencies, boards, and commissions must be examined as an integral part of public policy and of the organizational modes available to policy-makers. *(p. 158)*

To provide closer insight into the areas that are covered by government regulation in the field of policing and public safety, we will look at the scope of topics dealt with in several provincial jurisdictions. We will consider the use of the regulatory policy instruments in Nova Scotia, Ontario, and British Columbia.

Nova Scotia

In Nova Scotia, the *Police Act* (R.S.N.S., c. 348) includes regulations that deal with the following areas of activity:

- public complaints, internal discipline, and suspension;
- Police Review Board;
- municipal police ranks, insignia, service badges, and clothing; and
- police training.

Ontario

In Ontario, the *Police Services Act* (R.S.O. 1990, c.P.15) contains regulations dealing with the following aspects of policing:

- adequacy and effectiveness of police services;
- conduct and duties of police officers respecting investigations by the Special Investigations Unit;
- equipment and use of force;
- code of conduct; and
- political activity of municipal police officers.

British Columbia

In British Columbia, the *Police Act* (R.S.B.C., c. 376) includes regulations on the following areas:

- Organized Crime Agency of British Columbia;
- police disposal of property;
- rules regarding training, certification, and registration of municipal constables;

- special provincial constable complaint procedure; and
- Stl'atl'imx Tribal Police Service operations.

POLICY REVIEW AND EVALUATION

The capacity to review and evaluate public policy issues is something that challenges all levels of government across Canada. As noted earlier in this chapter, many government departments include individuals and units capable of providing detailed policy planning and development capabilities. However, there is a growing network of organizations and agencies that are operating in conjunction with government departments to advance the public policy agenda. What follows is a brief outline of some of the key areas of external support that exist within the Canadian context to provide policy review and evaluation.

Institute for Research on Public Policy

Based in Montreal, the Institute for Research on Public Policy (IRPP) is a non-partisan and independent think-tank that seeks to promote innovative ideas that generate better public policy choices in Canada. The IRPP seeks to foster genuine dialogue and debate on important policy issues and to take a strategic approach to the challenges of democratic governance, economic stability, and social progress.

With a strong commitment to public policy research, the IRPP has adopted a mandate that features an interdisciplinary and integrated interest in a fundamental question: What choices must Canada make to optimize and integrate economic performance and social progress in the face of rapid global change, and how can these choices best be made?

The IRPP has an impressive range of publications that represent the organization's adherence to its research mandate. The following are among the significant aspects of this effort:

Policy Options.   This journal deals with relevant public policy issues and presents the work of academics, practitioners, and public sector experts.

Policy Matters.   In early 2000, the IRPP launched a series of working papers under this collective title to provoke debate on particular matters of significance for Canadian public policy.

Choices.   This series presents the original work of academics on major public policy concerns.

Research projects.   These research projects may be highly relevant to police and public safety organizations. In the area of governance, the IRPP is pursuing research projects on strengthening Canadian democracy; courts and legislatures; governing the Canadian federation; and creating a new social union.

Public Policy Forum

The Public Policy Forum was created in the late 1980s to provide a focus for the exploration of issues pertaining to government and the federal public service. The organizations that were involved at the beginning of the Public Policy Forum included academic, labour, business, government, media, and volunteer sector group.

Among the strategic themes being pursued by the Public Policy Forum, the following pertain directly to policing and public safety across Canada:

- trust and confidence in public institutions; and

- public sector management.

As well as sponsoring workshops and conferences that bring together individuals and institutions concerned about the quality of public policy, the Public Policy Forum publishes a quarterly newsletter, *Forum,* as well as publications and essays designed to address emerging public policy issues.

Queen's University, School of Policy Studies

This school offers a master's degree in public administration and provides students with a range of courses that are designed to explore and explicate theories and practices relevant to the study of public policy in Canada. The following listing provides a summary of some of the course topics offered through this school:

- governing institutions that are involved with policy problems;

- approaches to policy analysis, including research strategies, methods, and techniques;

- law and public policy, including an assessment of how common law and constitutional law affect the exercise of government authority;

- policy modeling and decision-making, including a range of analytical tools that assist in the formulation and assessment of public policy, such as simulations, constrained optimization, linear programming uncertainty, and game theory;

- conflict management, focused on its application in public sector organizations;

- public sector reform; and

- trade and public policy including issues associated with international trade, for example, the World Trade Organization.

Beyond the three organizations noted above, it is worthwhile to identify a few academic publications that carefully examine key public policy issues that may affect the Canadian scene:

- *Canadian Public Policy*

- *Canadian Public Administration*

- *Governance*

- *Journal of Public Policy*

- *Law and Economics*

- *Yale Journal on Regulation*

- *Policy and Politics*

- *Regulation*

- *Public Administration Review*

CONCLUSION

This chapter has presented some background information on the fundamentals of policy analysis and development. We have considered a couple of models that provide a framework for this area of study and have examined some of the components of these models.

Within the context of individual police departments in Canada, we have seen how policy communities have developed and how they facilitate the formulation of policy as it pertains to policing and public safety. Every police service in Canada must be prepared to perform those functions normally associated with the public policy cycle, as outlined in this chapter.

We have briefly considered some of the policy instruments that are available to decision-makers in government. Our examination has focused on the use of regulatory instruments in the area of policing; however, there are many other policy instruments available to government officials in responding to the demands of public safety and security.

Finally, we have considered some of the key organizations and institutions that function to support and extend the capacity of the various levels of government and bureaucracy in their response to the challenges of policy review and evaluation. Policy analysis and development cannot take place within a vacuum, and these independent agencies are essential to the public policy enterprise in Canada.

REFERENCES

Aucoin, Peter (1996). *The new public management: Canada in comparative perspective.* Montreal : Institute for Research on Public Policy.

Bardach, Eugene (2000). *A practical guide for policy analysis: the eightfold path to more effective problem solving.* New York : Chatham House Publishers.

Barberis, Peter (1998). "The new public management and a new accountability." *Public Administration,* Vol. 76 (Autumn), pp. 451–470.

Barrow, Michael (1996). "Public services and the theory of regulation." *Policy and Politics,* Vol. 24, No. 3 (July), pp. 263–276.

Baxter-Moore, Nicholas (1987). "Policy implementation and the role of the state: a revised approach to the study of policy instruments." In Jackson, Robert et al. (eds.). *Contemporary Canadian politics: readings and notes.* Scarborough, Ont. : Prentice-Hall Canada.

Bellamy, Christine and John A. Taylor (1998). *Governing in the information age.* London : Open University.

Brooks, Stephen (1993). *Public policy in Canada: an introduction. 2nd ed.* Toronto : McClelland & Stewart.

Clifford, John and Kernaghan Webb (1986). *Policy implementation, compliance and administrative law.* Ottawa : Law Reform Commission of Canada.

Courchene, Thomas (1990). "Toward the reintegration of social and economic policy." In Doern, G. Bruce and Bryne Purchase (eds.). *Canada at risk?* Toronto : C.D. Howe Institute.

Dobuzinskis, Laurent, Michael Howlett, and David Laycock (eds.) (1996). *Policy studies in Canada: the state of the art.* Toronto : University of Toronto Press.

Doern, G. Bruce (1979). "Regulatory processes and regulatory agencies." In Doern, G. Bruce and Peter Aucoin (eds.). *Public policy in Canada: organization, process, and management.* Toronto : Macmillan of Canada.

———— (1994). *The road to better public services: progress and constraints in five federal agencies.* Montreal : Institute for Research on Public Policy.

Doern, G. Bruce and V. Seymour Wilson (eds.) (1974). *Issues in Canadian public policy.* Toronto : Macmillan.

Doern, G. Bruce and Peter Aucoin (eds.) (1979). *Public policy in Canada: organization, process, and management.* Toronto : Macmillan of Canada.

Doern, G. Bruce and Richard Phidd (1983). *Public policy: ideas, structure, process.* Toronto : Methuen.

Doern, G. Bruce, Leslie Pal, and Brian Tomlin (eds.) (1996). *Border crossings: the internationalization of Canadian public policy.* Toronto : Oxford University Press.

Dunleavy, Patrick (1991). *Democracy, bureaucracy and public choice.* London : Harvester Wheatsheaf.

Dyck, Rand (1993). *Canadian politics: critical approaches.* Toronto : Nelson Canada.

Feick, Jurgen (1992). "Comparative policy studies: a path towards integration." *Journal of Public Policy,* Vol. 12, No. 3 (July–September), pp. 257–285.

Ferlie, Ewan et al. (1996). The new public management in action. Oxford : Oxford University Press.

Ford, Robin and David Zussman (1997). *Alternative service delivery: sharing governance in Canada.* Toronto : Institute of Public Administration of Canada.

Gruber, Judith E. (1988). *Controlling bureaucracies.* Berkeley : University of California Press.

Harris, Richard A. (1989). *The politics of regulatory change: a tale of two agencies.* Oxford : Oxford University Press.

Hart, Michael and Brian Tomlin (2002). "Inside the perimeter: the US policy agenda and its implications for Canada." In Doern, G. Bruce (ed.). *How Ottawa spends 2002–2003: the security aftermath and national priorities.* Toronto : Oxford University Press.

Hockin, Thomas A. (ed.) (1977). *Apex of power: the prime minister and political leadership in Canada. 2nd ed.* Scarborough, Ont. : Prentice-Hall of Canada.

Hood, Christopher (1996). "Control over bureaucracy: cultural theory and institutional variety." *Journal of Public Policy,* Vol. 15, No. 3, pp. 207–230.

Janis, Irving L. (1989). *Crucial decisions: leadership in policymaking and crisis management.* New York : Free Press.

Keck, Margaret E. and Kathryn Sikkink (1988). *Activists beyond borders: advocacy networks in international politics.* Ithaca : Cornell University Press.

Kernaghan, Kenneth (1993). "Reshaping government: the post-bureaucratic paradigm." *Canadian Public Administration,* Vol. 36, No. 4 (Winter), pp. 636–645.

———— (2000). *Rediscovering public service: recognizing the value of an essential institution.* Toronto : Institute of Public Administration of Canada.

Larsen, E. Nick (1996). "The effect of different police enforcement policies on the control of prostitution." *Canadian Public Policy,* Vol. 22, No. 1 (March), pp. 40–55.

Larson, Peter (2002). *Ten tough jobs.* Ottawa : Public Policy Forum.

Lindquist, Evert A. and James Desveaux (1998). *Recruitment and policy capacity in government: an independent research paper on strengthening government's policy community.* Ottawa : Public Policy Forum.

March, James G. and Johan P. Olson (1989). *Rediscovering institutions: the organizational basis of political life.* New York : Free Press.

Pal, Leslie (2001). *Beyond policy analysis: public issue management in turbulent times. 2nd ed.* Toronto : ITP Nelson.

Peters, B. Guy (2001). *The politics of bureaucracy. 5th ed.* London : Routledge.

Public Policy Forum (2002). *Learning and skill development: issues, best practices and suggestions for action: the views of leaders at seven regional workshops and a national workshop.* Ottawa : Public Policy Forum.

———— (2002). *Public infrastructure in Canada: status, priorities and planning: background report based on a series of interviews with governments, associations, investors and others.* Ottawa : Public Policy Forum.

Rice, James and Michael Prince (2000). *Changing politics of Canadian social policy.* Toronto : University of Toronto Press.

Schellenberg, Kathryn (1997). "Police information systems, information practices and personal privacy." *Canadian Public Policy,* Vol. 23, No. 1 (March), pp. 23–39.

Schon, D. and M. Rein (1994). *Frame reflection: toward the resolution of intractable policy controversies.* New York : Basic Books.

Seidle, Leslie (1996). *Rethinking the delivery of public services to citizens.* Montreal : Institute for Research on Public Policy.

Smith, Jennifer and Susan Snider (1998). *Facing the challenge: recruiting the next generation of university graduates to public service.* Ottawa : Public Service Commission of Canada and the Public Policy Forum.

Sossin, Lorne (1993). "The politics of discretion: toward a critical theory of public administration." *Canadian Public Administration,* Vol. 36, No. 3 (Fall), pp. 364–391.

Trebilcock, Michael J. et al. (1982). *The choice of governing instrument: a study prepared for the Economic Council of Canada.* Ottawa : Canadian Government Publishing Centre.

Tupper, Allan and G. Bruce Doern (1981). *Public corporations and public policy in Canada.* Montreal : Institute for Research in Public Policy.

Wilson, V. Seymour (1973). "The role of royal commissions and task forces." In Doern, G. Bruce and Peter Aucoin (eds.). *The structures of policy-making in Canada.* Toronto : Macmillan of Canada.

WEBLINKS

Canadian Public Policy

http://economics.ca/cpp/en

Institute for Research on Public Policy

www.irpp.org

Public Policy Forum

www.ppforum.com/english

Queen's University, School of Policy Studies

http://qsilver.queensu.ca/sps

QUESTIONS FOR CONSIDERATION AND DISCUSSION

1. What policy instruments are most appropriate for guiding public safety and policing activities in Canada?

   _____

   _____

   _____

   _____

   _____

   _____

   _____

   _____

   _____

   _____

2. Which phase of the public policy cycle requires the most input from police departments in Canada?

   _____

   _____

   _____

   _____

   _____

   _____

   _____

   _____

   _____

   _____

3. What role, if any, should think-tanks play in the development of public policy as it relates to policing and public safety?

_____

_____

_____

_____

_____

_____

_____

4. What methods could be used to evaluate policies pertaining to police operations in Canada?

_____

_____

_____

_____

_____

_____

_____

_____

RELATED ACTIVITIES

1. Review the research agenda for any of the following organizations and discuss its possible relationship with policing and public safety issues:

   • Institute for Research on Public Policy

   • Institute of Public Administration of Canada

   • Public Policy Forum

2. Examine the various provincial statutes dealing with police and compare the scope and specifics of the regulations in each jurisdiction. Highlight the similarities and differences among these regulatory frameworks.

3. Select an important policing or public safety policy issue and trace its evolution following the public policy cycle outlined in this chapter.

4. Identify an emerging issue in the field of policing and apply Bardach's eightfold approach to policy analysis.

5. Examine current and back issues of the academic journals identified in this chapter for relevant articles on topics relating to policing and public safety.

6. Explore the material provided by the federal government's Policy Research Initiative (PRI). The website for this initiative, at **www.policyresearch.gc.ca**, provides a wealth of information pertaining to a wide range of research that serves to inform the development of public policy. The PRI explores ways of building capacity to support the formulation of public policy and ensuring that various levels of government have the necessary capabilities to support the range of public policy functions. The PRI also sponsors a Learning Advisory Panel on Policy which is intended to address the needs of professionals within the policy community. This panel has undertaken several initiatives in support of its mandate, including the formulation of a competency profile for policy practitioners.

chapter eight

Canadian Government Bureaucracy

For Themistocles was a man in whom most truly was manifested the strength of natural judgement, wherein he had something worthy [of] admiration different from other men. For by his natural prudence, without the help of instruction, before or after, he was of extemporary matters upon short deliberation the best discerner, and also of what for the most part would be their issue, the best conjecturer. What he was perfect in, he was able also to explicate: and what he was unpractised in, he was not to seek how to judge of conveniently. Also, he foresaw, no man better, what was the best or worst in any case that was doubtful. And (to say all in few words) this man, by the natural goodness of his wit, and quickness of deliberation, was the ablest of all men to tell what was fit to be done upon a sudden. But falling sick he ended his life: some say, he died voluntarily by poison, because he thought himself unable to perform what he had promised the king.

—Thucydides, *History of the Peloponnesian War* (translated by Thomas Hobbes)

Royal Newfoundland Constabulary.

Newfoundland Constabulary Mounted Division, circa 1890.

Learning Objectives

1. Identify the nine key values outlined for the Canadian public service.
2. Understand the importance of bureaucratic structures in federal, provincial, territorial, regional, and municipal governments in Canada.
3. Identify the various public service commissions across Canada and list their core functions.
4. Describe the importance of controlling the public service bureaucracy in Canada.
5. Identify several elements of bureaucratic dysfunction.

INTRODUCTION

Across all levels of government in Canada there is some form of bureaucracy that is responsible for carrying on the day-to-day administration and operation of the work of public service. This includes the vast range of activities carried out through the federal civil service as well as the myriad of services provided through your local municipality. Each of these levels must engage individuals who not only have some genuine commitment to the values and principles of public service, but must also be in possession of the specific knowledge, skills, and abilities to perform tasks that are essential to services being provided by that level of government.

This chapter will consider the following three main levels of bureaucracy:

- federal

- provincial or territorial

- municipal or regional

It will quickly become apparent that many of the fundamental features of the public service bureaucracy are shared across all levels of government. This allows us to focus on some important shared values and to examine key elements that are endemic to public services in every jurisdiction of Canada.

Finally, we will look at the ways of dealing with bureaucratic dysfunction and the efforts being made in several jurisdictions to engage in substantial public service renewal.

FEDERAL BUREAUCRACY

The Public Service Commission of Canada (PSC) is the organization responsible for the overall guidance of the federal civil service. It is an independent agency that has responsibility for ensuring that the values of the Canadian public service are maintained. It does this largely through its role in the administration of the *Public Service Employment Act,* as well as through the application of a merit-based system of recruitment, appointment, and promotion of federal public servants. It delivers training, education, and development programs across the federal civil service and observes the requirements of the *Employment Equity Act.*

The Public Service Commission of Canada presently comprises the following five branches:

- *Staffing and Recruitment Programs Branch.* This branch works with all federal departments and agencies in order to ensure that the Canadian taxpayer has a highly proficient public service that is representative of our society. It has responsibility for operational policy and program design, including the efficient delivery of products and services through the PSC's regional and district offices. As well as being responsible for public service recruitment, this branch ensures the delivery of employment equity initiatives and several corporate programs on behalf of the Treasury Board.

- *Learning, Assessment and Executive Programs Branch.* This branch is responsible for meeting the current and future staffing needs of the federal government at the executive level. The branch offers a range of learning programs designed to improve job-related skills. The branch is also involved in the identification of core competencies required by senior public servants, as well as the development of selection tools and an array of learning products and services.

- *Policy, Research and Communications Programs Branch.*   This branch supports the overall policy goals of the PSC and works to ensure that Canada has an independent, professional, and representative civil service. The branch functions to advance medium- and long-range planning on behalf of the PSC and conducts strategic research and analysis, as well as environmental scanning.

- *Recourse Branch.*   This branch is mandated to provide suitable processes that support the merit principle in the Canadian public service. By means of effective intervention and education, the branch works to ensure the genuine independence of the quasi-judicial appeals and investigations that take place in the PSC. This branch will conduct hearings into alleged breaches of the *Public Service Employment Act* and its regulations. It also provides training, guidance, and support to federal departments, unions, and other organizations or individuals, as appropriate.

- *Corporate Management Branch.*   This branch provides systems and services in support of the executive management of the PSC's programs and activities, including management systems, financial services, human resources management, internal audit, and information systems, among others.

OUTLINE OF PUBLIC SERVICE VALUES

Within the federal public service there has been a great deal of discussion and dialogue around the appropriate values that should inform the work of the civil service. To that end, the following public service values have recently been articulated by the Public Service Commission of Canada:

- *Merit.*   Recruitment and promotion should be based on skills and qualifications appropriate to the job, not on favouritism or political affiliation;

- *Political neutrality.*   Public servants must serve the government of the day and speak truth to power. Public servants are free to vote, but should not bring their political beliefs to the workplace.

- *Accountability.*   Originally interpreted as administrative accountability, concern for the way in which things are done; today this value is increasingly focused on the outcome of actions.

- *Anonymity.*   This convention is closely bound to political neutrality and ministerial responsibility. Politicians protect the anonymity of public servants in exchange for their neutral advice. Public servants cannot speak in the House and so rely on ministers to defend their actions. This convention is under threat for different reasons; for example, public servants are increasingly in the public eye through public consultations and are encouraged to speak to media. There have been cases where public servants were publicly named by their ministers without an opportunity to clear their reputations (Sutherland, 1991);

- *Economy, efficiency, and effectiveness.*   These three watchwords for the conduct of public servants tie into the concept of public servants being stewards of taxpayers' money and guardians of the public interest.

- *Responsiveness.* Public servants are expected to be sensitive to political and public needs.

- *Representativeness.* The public service should reflect the composition of the population at large in terms of gender, religious, ethnic, and socioeconomic groups.

- *Fairness and equity.* Public servants should deliver their service to citizens without special favour.

- *Integrity.* Closely tied to honesty and judgement, integrity is fundamental to nurturing public trust in government. The d ment and dissemination of clear conflict of interest guidelines is one means of encouraging integrity. (Source: Institute on Governance. See the Public Service Commission of Canada website, **http://www. edu.psc-cfp.gc.ca/tdc/learn-apprend/psw/hgw/valeurs_e.htm**)

PROVINCIAL AND TERRITORIAL BUREAUCRACIES

As noted earlier, every provincial and territorial jurisdiction in Canada has a formally organized civil service infrastructure that has responsibility for the delivery of a wide range of public services. This chapter will not provide details on every single jurisdiction; however, the information provided below should be seen as highly representative of the types of programs and services offered in all jurisdictions in Canada. Each provincial and territorial government website offers current information to readers considering the impact of bureaucratic structures on policing and public safety services across Canada.

Newfoundland and Labrador

In this province, the Public Service Commission has the mandate to ensure that a politically neutral and professional public service is essential for maintaining the style of parliamentary democracy enjoyed within that jurisdiction. The commission came into existence with the enactment of the *Public Service Commission Act* in 1974. The commission is empowered to protect the merit principle in the appointment and promotion of civil servants across the entire range of government departments and agencies.

Accordingly, the commission has committed itself to the following values:

- public service excellence;

- respect for the individual; and

- merit, fairness, and transparency in public service work.

Prince Edward Island

Within the Provincial Treasury department is PEI's Public Service Commission. The commission is an independent agency that holds responsibility for providing coordination and leadership for the human resources within the province's public sector. It serves all other provincial government departments and agencies, as well as the five regional health authorities. It offers a comprehensive range of human resources services, including classification, recruitment and selection, and related functions.

Nova Scotia

The Nova Scotia Public Service Commission is tasked with four significant functions:

- developing and implementing corporate human resources policy, programs, and services;
- ensuring the quality and value of human resources management practices through a process of audits and evaluations;
- ensuring fair and effective hiring practices within the public service; and
- acting as the government's agent for collective bargaining.

The Public Service Commission was created by an act of the provincial legislature in June 2001. It is staffed by nearly 65 civil servants with skills and abilities in a range of activities commensurate with the commission's mandate.

New Brunswick

In New Brunswick, the Office of Human Resources serves the needs of the provincial government in a manner similar to a public service commission. The organizational chart of this department is presented in Figure 8.1.

Ontario

The Ontario Public Service Excellence and Innovation Office was created recently within the Cabinet Office to facilitate changes in the focus of the Ontario Public Service (OPS).

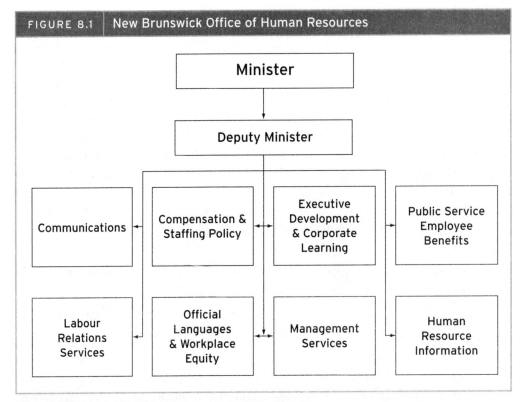

FIGURE 8.1   New Brunswick Office of Human Resources

Source: New Brunswick Office of Human Resources. Used with permission.

The vision of this new entity is a governmental culture that promotes better outcomes and significant service improvements for its customers and clients. A great deal of work has been invested in the formulation of a series of "common service standards" which can be applied across the OPS and which will, in fact, constitute a framework for action within the civil service. As part of its overall goals for the improvement of service quality and effectiveness, the OPS aims to

- increase public satisfaction with its services;

- ensure that Ontario measures up to external benchmarks within the public and private sectors; and

- ensure that Ontario is seen as a benchmark for quality services by similar jurisdictions.

The OPS, in adopting a consistent set of principles for its efforts, asserts that quality in public service

- must be focused on customers and citizens;

- is everyone's business;

- must be recognized and valued;

- must be measured against standards;

- must respect the diversity of OPS businesses, customers, and stakeholders;

- should be achieved through ongoing learning and improvement; and

- requires sustained commitment and leadership.

As an additional focus for its activities, the OPS has articulated six "cornerstones" of quality service, presented here in Figure 8.2.

Saskatchewan

The Saskatchewan Public Service Commission is an independent body that holds responsibility for providing effective human resources management within this province. It also represents the public interest through its administration of *The Public Service Act, 1998*. Its work includes recruiting and selecting suitable candidates for public service positions across the government, developing position classification standards, representing management with respect to employee relations, and related activities.

In developing detailed staffing standards, the Saskatchewan Public Service Commission has applied a useful competency model that ensures a focus on the following components:

- *Personal attributes.*   These are the underlying characteristics that define an individual's personal style and effectiveness.

- *Skills.*   These are capacities that may be observed and will be essential to certain public service positions. Skills may be developed or enhanced through practice, learning, and experience.

- *Knowledge.*   This relates to the fundamental information that gives a person the wherewithal to perform job tasks and function in an intelligent manner. Knowledge

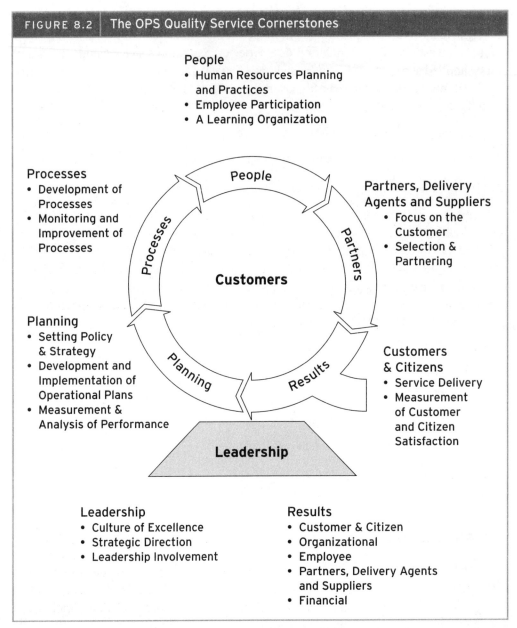

FIGURE 8.2 | The OPS Quality Service Cornerstones

People
- Human Resources Planning and Practices
- Employee Participation
- A Learning Organization

Processes
- Development of Processes
- Monitoring and Improvement of Processes

Partners, Delivery Agents and Suppliers
- Focus on the Customer
- Selection & Partnering

Planning
- Setting Policy & Strategy
- Development and Implementation of Operational Plans
- Measurement & Analysis of Performance

Customers & Citizens
- Service Delivery
- Measurement of Customer and Citizen Satisfaction

Leadership
- Culture of Excellence
- Strategic Direction
- Leadership Involvement

Results
- Customer & Citizen
- Organizational
- Employee
- Partners, Delivery Agents and Suppliers
- Financial

Source: Government of Ontario website. Used with permission.

may include possession of theories, facts, and principles pertaining to a particular area of activity and may be acquired through formal or informal learning experiences.

Alberta

In Alberta, the largest portion of the executive branch of government is the public service. In 2002, the Alberta public service included more than 22,000 people located in nearly 160 different centres across the province. The Personnel Administration Office within the Alberta government has oversight of the Alberta Public Service. It is responsible for the development of a professional civil service in the province and ensures that in the pursuit of its mission to provide quality government services to Albertans, it is guided by the following six fundamental principles:

- meeting Albertans' needs;

- providing relevant government services;

- increasing our partnerships;

- seeking innovative approaches;

- demonstrating fiscal responsibility; and

- demonstrating effectiveness.

British Columbia

In September 2001, British Columbia began its Public Service Renewal Project. The project's mandate was to "rebuild and sustain a professional public service that provides quality services to meet the needs of British Columbians."

By simplifying human resources management practices and fostering a supportive environment within the workplace, this project aims to achieve a level of excellence in the provision of public service. To meet the government's overall strategic direction, this renewal project has the following six goals:

- effective people strategy;

- proactive and visionary leadership;

- performance-focused workplace;

- flexible and motivating work environment;

- learning and innovative organization; and

- progressive employee-employer relations.

Yukon

The Yukon Public Service Commission was created by the *Public Service Act* to provide an extensive range of human resources and support services for all departments and employees within Yukon. Accordingly, the commission delivers recruitment, position classification, employee compensation and benefits, employee relations, and public service training and development programs.

The Yukon Public Service Commission comprises the following branches:

- *Corporate Human Resource Services,* which works to achieve the human resources management objectives of the various government departments in areas such as organization and job design, job evaluation, staffing, employment equity initiatives, First Nation recruitment practices, and other related areas;

- *Staff Development Branch,* which provides programs and services that help employees and departmental managers develop the knowledge, skills, and abilities required for effective organizational development and to address personal challenges that may have an impact on work performance.

- *Pay and Benefits Management,* which is responsible for the regular process of payroll administration and the management of the approved range of employee benefits.

- *Staff Relations Branch,* which is responsible for providing advice and assistance to public servants, government departments, and crown corporations regarding the interpretation and application of collective agreements and other governing instruments relating to employment. This branch is also engaged in collective bargaining on behalf of the Government of Yukon, and grievances and appeals under collective agreements, the *Public Service Act,* the *Education Act,* and the *Public Service Staff Relations Act.* Finally, this branch works closely with the Department of Indian and Northern Affairs in the devolution of the Northern Affairs Program (NAP) with the federal government.

MUNICIPAL AND REGIONAL BUREAUCRACIES

Across Canada there are municipal and regional governments that have a vital interest in the delivery of local policing and public safety services. It is, therefore, important to understand the dynamic relationship that exists between municipal or regional police departments and their local government counterparts. Several detailed discussions of the governance model that pertain to various Canadian municipalities are available for further study (Stenning, 1981; Hann et al., 1985; McKenna, 2002).

Halifax Regional Municipality

The Halifax Regional Municipality is governed by a council consisting of the mayor and 23 municipal councillors, with each councillor representing a specific district within the municipality. The mayor and councillors sit on various committees of council and represent the legislative body for the municipality. The Chief of Police for the Halifax Regional Police Service has a reporting relationship within the bureaucratic structure of the regional municipality. Accordingly, the chief is accountable to the Halifax Regional Council, the Chief Administrative Officer, and the Police Commission. Figure 8.3 provides an overview of the Halifax Regional Municipality and shows the interrelationships that exist within this bureaucratic structure.

As part of its vital interest in matters pertaining to policing and public safety, the Halifax Regional Municipality commissioned a detailed review of policing services within its jurisdiction. This study examined the operation of the Halifax Regional Police and the Halifax Detachment of the RCMP. The team assembled to undertake this study included individuals with academic, police, and private industry experience. The project included a review of existing literature, extensive interviews, focus groups, and on-site inspections of police facilities and operations. Among the recommendations presented by the consultants were the following:

- improved business planning;
- shared responsibility with the community; and
- increased efficiency and effectiveness of service delivery.

Toronto

The Toronto Police Service is funded by the taxpayers of Toronto and has a reporting relationship to the Toronto Police Services Board. Currently, the board comprises seven members appointed as follows:

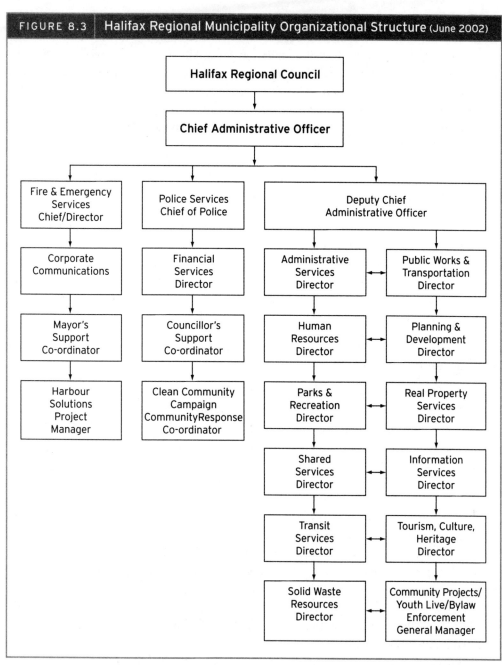

FIGURE 8.3 | Halifax Regional Municipality Organizational Structure (June 2002)

Source: Halifax Regional Municipality. Used with permission.

- the head of Toronto City Council, or another member appointed by council;

- two members of council appointed by resolution of council;

- one person appointed by resolution of council who is not a member of council; and

- three persons who are appointed by the Province of Ontario (i.e., the Lieutenant Governor in Council).

While a substantial amount of activity relating to policing and public safety falls within the bureaucratic control of the Toronto Police Service, there are other structures and systems in place within the municipality that concern themselves with these issues.

For example, Toronto City Council sponsors a Task Force on Community Safety. In responding to public concerns about crime and disorder, this task force was established by council in January 1998. It was intended to formulate a comprehensive and coordinated plan that would demonstrate Toronto's leadership in the area of community-based crime prevention. The task force includes approximately 22 members who represent key municipal stakeholder groups, including the Toronto Police Service, educators, neighbourhood groups, local businesses, and agencies and organizations dedicated to the prevention of violence. In September 2000 the task force tabled a major report, entitled *Toronto's community safety strategy: turning words into action to make Toronto safer.*

Also, Toronto City Council has initiated the Woman Abuse Council to facilitate the development of a coordinated, community-based response to violence against women. Membership in this group includes the Toronto Police Service, local shelters, support service agencies, hospitals, community health facilities, probation personnel, and survivors of abuse. The Woman Abuse Council has undertaken several initiatives directed at this significant public safety concern.

Winnipeg

The departments and major organizational units that make up the City of Winnipeg's infrastructure are as follows:

- Audit
- CAO Secretariat
- Chief Financial Officer
- City Clerk's Department
- Community Services
- Corporate Information Technology
- Corporate Finance
- Corporate Services
- Emergency Preparedness Program
- Film and Cultural Affairs
- Fire Paramedic Service
- Office of the Ombudsman
- Mayor's Office
- Planning, Property and Development
- Public Works
- Water and Waste
- Winnipeg Hydro
- Winnipeg Police Service
- Winnipeg Transit

In addition to these bureaucratic elements within the infrastructure of the municipality, there exists the Winnipeg Committee for Safety (WCFS). The WCFS was created by City Council in 1999 and is intended to facilitate and support the efforts of the local community in addressing safety concerns. The WCFS reports to the Executive Policy Committee of City Council and serves to function as a link between the community and council. Representatives on the WCFS are taken from the following layers of bureaucracy:

- Civic Administration (including Community Services)

- Winnipeg Police Service

- Chief Administrative Office Secretariat

- City Clerk's Department (non-voting membership)

- Department of Justice (provincial representatives)

The WCFS has been active in several community activities, and its efforts have resulted in some of the following accomplishments:

- *Community Conversations Manual,* a guide for planning community safety discussions;

- *Crime Prevention Manual,* a component of a *Safety Tool Box* that consolidates relevant documents and materials for a safe city;

- *Winnipeg Safety & Security Guide Call Card 2001/2002,* a listing of agencies and individuals that may be contacted with regard to community safety and security issues; and

- *Safety Audit Manual,* another component of the *Safety Tool Box,* designed to assist in the assessment and improvement of safety in the community.

Vancouver

The Vancouver Police Department is under the authority of the Vancouver Police Board, which is created and empowered by the *British Columbia Police Act.* As with other municipal police services, the Vancouver Police Department receives significant funding from the City of Vancouver, and the mayor currently chairs the Vancouver Police Board. Recently, the board initiated a search process to replace the outgoing chief of police. This, of course, is an important decision for any police governing authority and requires an enormous attention to detail. Also, the board must consider the key priorities and challenges that a new chief of police will face upon selection. Accordingly, among the priorities and challenges for the first year of the new chief's tenure identified by the Vancouver Police Board are the following:

- Work with the city to ensure that SAP, the recently installed e-business system, generates relevant, timely information to permit informed decisions regarding resource allocation.

- Undertake a thorough organizational review, with a focus on structure and allocation of responsibilities, workloads, and skill requirements. The guiding principles of this review will be improved services to the community and more effective management of the organization and its resources.

- Develop an open, constructive relationship with colleagues in the city government — contributing to discussion and decisions regarding broader issues affecting the city, while helping colleagues understand the special needs and challenges of the police department.

- Establish a strong working relationship with government and community partners involved in the Vancouver Agreement and the city's Four-Pillar Approach to Drug Problems in Vancouver.

- Develop a comprehensive and coordinated enforcement and policing plan, and a policy framework to address drug issues and the problems in the Downtown Eastside. (Source: Vancouver Police Department. *Chief Constable briefing document.* March 7, 2002.)

The Vancouver Police Department is also involved in the work of the city's Neighbourhood Integrated Service Teams (NIST). These teams were adopted by a resolution of the Vancouver City Council on September 27, 1994, and are made up of city and other community agency representatives who work across organizational and bureaucratic boundaries to address community problems. The overall goals of the NIST include the following:

- to create safer and more pleasant neighbourhoods;

- to provide effective and friendly services;

- to facilitate community involvement in creative problem-solving; and

- to provide ready access to information about the city and its government.

CONTROLLING BUREAUCRACY

In recognition of the power of the bureaucracy to influence the lives of Canadians at various levels of activity, it is important that there be mechanisms in place to control the public service and to ensure its accountability to the public interest.

At the apex of controls is the federal Cabinet and those other provincial, territorial, regional, or municipal bodies that represent the central decision-making authority within any given jurisdiction. Additionally, within individual ministries or departments, the key elected official should be someone whose role involves the control of the department's bureaucratic structure. While the reality may not perfectly reflect the theory in this regard, it is intended that ministerial controls will be in place. Certainly, government ministers are often highly dependent upon the guidance and advice of their bureaucrats, particularly when they are new to, or unfamiliar with, the complexities and nuances of their portfolio. However, at the Cabinet, or central agency, level there is often a highly developed capacity to assess and evaluate bureaucratic advice from a pragmatic, political perspective that provides an appropriate corrective to poor public service proposals.

Furthermore, there are in most Canadian bureaucracies certain checks and balances across bureaucratic structures. For example, at the federal level, the Public Service Commission of Canada has a degree of oversight over human resources issues pertaining to all federal departments. Auditors General, at the federal and provincial levels, have a broad capacity to review, assess, and pass judgement on the efficiency and effectiveness of any government department or agency within their jurisdiction. This amounts to a significant bureaucratic control. Within the government of Ontario, the Management Board of Cabinet has extensive control over the policies, procedures, programs, and priorities that are undertaken in every provincial ministry.

Also, as can be seen during the regular sessions of question period in Parliament, provincial legislatures, and municipal councils, elected officials often function to call the bureaucracy to account for their actions, or inaction. Beyond approving the expenditure of funds that come from the public purse, elected officials have an important role to play in calling upon the appropriate minister to explain or justify the functioning of the department in question, including the performance of the bureaucracy. Frequently, elected officials will work on behalf of their constituents to expedite matters of concern within the bureaucracy. This type of action may take the form of legislative change, or may merely relate to the securing of a government document or authorization that has been unduly delayed or inappropriately denied.

The judiciary in Canada also plays a significant role in controlling bureaucracy. Judges are, in many instances, empowered to overturn bureaucratic decisions if the decisions are determined to breach some fundamental constitutional or procedural rule of law. Certainly in this area of bureaucratic control, the *Canadian Charter of Rights and Freedoms* is pre-eminent and frequently touches directly upon those activities of police organizations that may be construed as being within the bureaucratic realm of public service (McKenna, 2002). Certainly, the important recent decisions in cases such as *R. v. Feeney* and *R. v. Stinchcombe* are in the category of control of the police as a special kind of bureaucratic operative.

There is the additional feature of the office of the Ombudsman within the Canadian bureaucratic system. In nearly all provinces, there is such an office that functions on behalf of the public good to ensure that no one is treated unfairly, improperly, or unreasonably by governments or their agents. What follows are some highlights of the work done by these offices in certain Canadian jurisdictions.

Bureaucratic Dysfunction

As with every organizational structure that relies upon human activity, public service bureaucracies are prone to error, accident, failure, and other forms of dysfunction. Various forms of constraint affect the capacity of any human organization to perform flawlessly (Janis, 1989). Extensive application delays, the frustrations associated with government "red tape," bureaucratic inaction, and failures in communications are all symptomatic of government processes that appear to be highly dysfunctional. These persistent realities tend to affect the public satisfaction with government services, including those that pertain to policing and public safety.

The "new" public management approach referred to in Chapter 6 ("The Origins of Public Administration") is one means of addressing possible sources of bureaucratic dysfunction. In Chapter 10 ("The Future of the Canadian Political Realm and Public Administration"), we will consider some of the ways in which bureaucratic dysfunction may be resolved.

CONCLUSION

This chapter has considered some of the key elements of the various levels of government bureaucracy. It should be clear that there are many patterns and common themes across the full range of bureaucratic structures that exist in Canada. It should also be clear that policing and public safety as currently delivered within the Canadian context involves bureaucratic systems and services that are highly aligned with government entities. At the federal, provincial, territorial, regional, and municipal levels of public administration, we find policing and public safety programs embedded in the bureaucratic infrastructure.

This chapter has also attempted to outline some of the major approaches for controlling the bureaucracy and the institutions and organizations that have been established to effect that control. Additionally, we have touched upon the challenge of bureaucratic dysfunction and will return to this theme later in the textbook.

REFERENCES

Armstrong, Jim (1997). *Stewardship and the public service: a discussion paper*. Ottawa : Public Service Commission of Canada.

Campbell, Colin (1983). *Governments under stress: political executives and key bureaucrats in Washington, London, and Ottawa*. Toronto : University of Toronto Press.

Campbell, Colin and George Szablowski (1979). *The superbureaucrats*. Toronto : Macmillan of Canada.

Granatstein, J.L. (1982). *The Ottawa men: the civil service mandarins, 1935–1957*. Toronto : Oxford University Press.

Hann, Robert G. et al. (1985). "Municipal police governance and accountability in Canada: an empirical study." *Canadian Police College Journal,* Vol. 9, No. 1, pp. 1–85.

Hodgetts, J.E. (1964). "Challenge and response: a retrospective view of the Public Service of Canada." *Canadian Public Administration,* Vol. 7, No. 4 (December), pp. 409–421.

———— (1973). *The Canadian public service: a physiology of government, 1867–1970*. Toronto : University of Toronto Press.

Janis, Irving L. (1989). *Crucial decisions: leadership in policymaking and crisis management*. New York : The Free Press.

Kernaghan, Kenneth (1991). "Career public service 2000: road to renewal or impractical vision?" *Canadian Public Administration,* (Winter), pp. 551–572.

———— (1999). *Rediscovering public service: recognizing the value of an essential institution*. Toronto : Institute of Public Administration of Canada.

Kernaghan, Kenneth and John Langford (1990). *The responsible public servant*. Halifax : Institute for Research on Public Policy.

McKenna, Paul F. (2002). *Police powers I*. Toronto : Pearson Education Canada.

———— (2003). *Police powers II*. Toronto : Pearson Education Canada.

Public Service Commission of Canada (1997). *The Public Service Commission of Canada at a glance*. Ottawa : Public Service Commission of Canada, Treasury Board Secretariat, and Privy Council Office.

Rowat, Donald (1968). *The ombudsman. 2nd ed.* Toronto : University of Toronto Press.

Stenning, Philip C. (1981). *Police commissions and boards in Canada*. Toronto : Centre of Criminology, University of Toronto.

Sutherland, S.L. (1991). "The Al-Mashat affair: administrative responsibility in parliamentary institutions." *Canadian Public Administration,* (Winter), pp. 573–603.

WEBLINKS

Alberta Office of the Ombudsman
www.ombudsman.ab.ca

Alberta Public Service Personnel Administration Office
www.pao.gov.ab.ca

British Columbia Ombudsman
www.ombud.gov.bc.ca

British Columbia Public Service Renewal
www.renewal.gov.bc.ca/overview.htm

Canadian Centre for Management Development
www.ccmd-ccg.gc.ca

Manitoba Ombudsman
www.ombudsman.mb.ca

Newfoundland and Labrador Public Service Commission
www.gov.nf.ca/psc

Nova Scotia Office of the Ombudsman
www.gov.ns.ca/ombu

Nova Scotia Public Service Commission
www.gov.ns.ca/psc

Office of the Commissioner of Official Languages
www.ocol-clo.gc.ca

Ombudsman at Canada Post
www.ombudsman.poste-canada-post.com

Ontario Ombudsman
www.ombudsman.on.ca

Ontario Public Service Excellence and Innovation Office
www.ontariodelivers.gov.on.ca

Prince Edward Island Public Service Commission
www.gov.pe.ca/pt/psc-info

Privacy Commissioner of Canada
www.privcom.gc.ca

Public Service Commission of Canada
www.psc-cfp.gc.ca

Quebec Ombudsman
www.ombuds.gouv.qc.ca/en

Saskatchewan Ombudsman
www.legassembly.sk.ca/officers/ombuds.htm

Saskatchewan Public Service Commission
www/gov.sk.ca/psc/about.htm

Treasury Board of Canada Secretariat Leadership Network
http://leadership.gc.ca

Yukon Public Service Commission
www.psc.gov.yk.ca

QUESTIONS FOR CONSIDERATION AND DISCUSSION

1. How important is the role of the Ombudsman to policing and public safety in Canada?

_____

_____

_____

_____

_____

_____

_____

2. Which values of the federal public service are most applicable to policing and public safety in Canada?

_____

_____

_____

_____

_____

_____

_____

3. Of the various approaches for controlling bureaucracy described in this chapter, which ones are most likely to influence the operational and administrative management of Canadian police services?

_____

_____

_____

_____

_____

_____

_____

_____

4. Which level of bureaucracy is most engaged with the delivery of policing services? Explain your answer.

_____

_____

_____

_____

_____

_____

_____

_____

RELATED ACTIVITIES

1. Study the work of the Ombudsman offices in various provincial jurisdictions and consider the impact their work might have upon different levels of policing in those jurisdictions.

2. Review the websites for the public service commissions in various Canadian jurisdictions and discuss their relationship with police services and public safety organizations.

3. Study some of the public service renewal projects that are currently being undertaken across Canada. Consider how police organizations are involved in these undertakings and make comparisons across different levels of government. For example, compare the innovations in the federal civil service with similar activities at a provincial or municipal level.

chapter nine

Public Policing—Critical Issues in Safety, Security, and Surveillance

Security is an essential need of the soul. Security means that the soul is not under the weight of fear or terror, except as a result of an accidental conjunction of circumstances and for brief and exceptional periods. Fear and terror, as permanent states of the soul, are well-nigh mortal poisons, whether they be caused by the threat of unemployment, police persecution, the presence of a foreign conqueror, the probability of invasion, or any other calamity which seems too much for human strength to bear.

—Simone Weil, *The need for roots: prelude to a declaration of duties towards mankind*

Royal Newfoundland Constabulary.

Deputy Chief Gary F. Browne, Royal Newfoundland Constabulary, receiving the first Member of the Order of Merit of the Police Forces from Her Excellency the Right Honourable Adrienne Clarkson, Governor General of Canada, 2002.

Learning Objectives

1. Identify key aspects relating to the challenge of terrorism for Canadian police organizations.
2. Identify major Canadian police and public agencies charged with the control of terrorism.
3. Identify key aspects relating to the challenge of organized crime for Canadian police organizations.
4. Identify major Canadian police and public agencies charged with the control of organized crime.
5. List critical aspects relating to computer crime in Canada.

6. Identify the key aspects relating to the challenge of information and communications technology for Canadian police organizations.

7. List key aspects of the debate over public versus private policing in Canada.

8. Identify the key aspects of the challenges facing Canadian police organizations with respect to protection of privacy and freedom of information.

9. Identify significant innovations in the area of police educational reform in Canada.

10. List key aspects pertaining to police oversight and civilian governance in Canada.

INTRODUCTION

This chapter will consider some of the most important aspects facing modern public policing in Canada. It will include an examination of the issues of safety, security, and surveillance that currently require the careful and constant attention of police organizations across the country. In doing so, we will attempt to better understand the leading concerns that police officers and police executives must grapple with in their day-to-day operations, as well as their long-range organizational planning. It is hoped that this chapter's focus on the most critical issues will make readers more alert to the depth of the challenges facing police services in Canada today.

We will begin with some review of the matter of terrorism and its immediate impact on policing in Canada. This will include some consideration of the area of police surveillance and will connect the need for such activity to the perceived threats deriving from national and domestic security concerns. This will be followed by some overview of the matter of computer crime as a special challenge for modern police departments, which will lead to a consideration of the whole question of information technology and its impact on the way in which police organizations must conduct their business.

We will move to a brief examination of the controversial issue of public versus private policing in Canada. This topic has generated a great deal of discussion and it is worthwhile summarizing some of the key elements of this debate. Next, we will look at matters relating to the protection of privacy (both individual and organizational) and the competing concerns for freedom of information. The delicate balance that is required in this area of legislation is critical to our fundamental rights and freedoms.

In completing our review of the major issues pertaining to public policing, we will look at the whole notion of educational reform. Policing has evolved significantly over the last few decades and it is clear that police organizations have been struggling to maintain consistent, conscientious, and competent programs for the ongoing training, education, and development of their members (both uniform and civilian). Accordingly, it is important to have a good understanding of the approaches that have been successful in this regard and the promising innovations that exist in this area of police leadership.

Finally, this chapter will touch upon the topic of police oversight and civilian governance. The way in which we, as a civilized society, guard the guardians is central to the quality of policing in Canada. By reviewing the quality and competence of our civilian governing authorities, we will quickly gain insight into the overall well-being of our policing institutions.

TERRORISM

September 11, 2001, will always be seen as a watershed in our collective lives as it pertains to immediate matters of safety and security. Very few events in our past history, outside of times of declared global war, have so profoundly disturbed and shaken our confidence in our ability to protect ourselves and our society from harm. While tragedy and calamity have been known to strike North Americans on other occasions, this event has proven to be a most sobering wake-up call with regard to the potential for terrorists to strike close to home and with devastating consequences.

Police organizations in Canada have an important role to play in the context of terrorism. They have been placed on a new level of alert to do whatever they can to support and consolidate the activities of other organizations with a mandate for fighting this form of crime, including the Canadian Security Intelligence Service (CSIS).

Police organizations spent much of the year following the September 11 attacks ensuring that they have in place protocols, procedures, policies, and contingency plans that prevent the threat of terrorism from compromising the safety and security of the Canadian public. All of the major police associations in Canada have focused attention on this issue during their annual conferences and have taken steps to consolidate their efforts to solidify a strategic and cross-jurisdictional approach to combating terrorism. For example, the Ontario Association of Chiefs of Police (OACP) worked hard to formulate a series of recommendations for the Ontario government. In October 2001, the OACP issued a document that contained key recommendations in the following areas pertaining to terrorism:

- *Staffing.*   Review staffing levels at all levels of policing in Ontario.

- *Training.*   In cooperation with the federal government, facilitate the development of adequate standards for a coordinated response to terrorism.

- *Technology.*   Ensure that police services are equipped with suitable and compatible equipment and technology to assist in counter-terrorist activities.

- *Financial requirements.*   Mount a coordinated effort involving all three levels of government to guarantee appropriate funding for policing in this area of activity.

- *Policy and legislation.*   Support federal legislative changes (e.g., Bill C-24 and Bill C-36) to combat terrorism and to ensure that immigration and refugee policies work to maintain national security.

ORGANIZED CRIME

The issue of organized crime has required a considerable amount of attention on the part of police services across Canada. Furthermore, governments at the federal and provincial levels have also had to turn their attention to this matter to introduce legislation that will facilitate the fight against organized crime by police organizations. The fact that organized crime is now a priority for law enforcement agencies attests to the success of police leaders in making the public aware of the dimensions of this public safety challenge. One aspect of organized crime that makes it possible for it to flourish in modern society is the fact that it frequently operates below the "radar" of regular citizens. When organized crime is most effective, it is least visible to those among us who are law-abiding citizens. It flourishes in circumstances where ordinary people are not aware of

its operations and have little or no contact with the activities of organized crime opera-
tives. Police leaders in Canada have invested a great deal of intelligence and effort in
making decision-makers aware of the magnitude and impact of organized crime.

When members of police organizations speak about organized crime, they typically
identify several distinct categories, or "genres," that must be dealt with in different
ways. The following listing provides a basic breakdown of these categories:

- Albanian

- Aboriginal

- Asian (including Chinese, Japanese, and Vietnamese gangs)

- Hispanic

- Italian

- Nigerian

- outlaw motorcycle gangs

- Russian and Eastern European

- youth gangs

Each of these categories requires specialized intelligence and a coordinated police
response that is unique to the group under consideration. For example, it is clear that
police forces must respond to outlaw motorcycle gangs in ways that are dramatically
different from their approach to Russian gangs or Asian triads. In many instances, the
linguistic skills required to effectively monitor the activities of various gang organizations present an important challenge to police organizations. Also, the close-knit nature
of organized gangs makes it exceedingly dangerous for those police officers willing to
conduct undercover work to gather intelligence and operationally sensitive insight that
will allow police to intervene when gangs are planning major illegal activities.

Among the activities that have been closely associated with organized crime are the
following:

- automobile theft

- corruption

- drug trafficking

- economic crime (including counterfeiting, fraud, and technological crimes)

- environmental crimes

- gambling

- influence peddling

- labour racketeering

- money laundering

- prostitution

- smuggling

When the police have gained a degree of information on the threat posed by organized
crime, they are typically quick to respond. However, the gathering and substantiating of
intelligence often requires an enormous investment of police resources to gather and sub-

stantiate intelligence. Accordingly, the work of Criminal Intelligence Service Canada is critical to the fight against organized crime. This organization's goals, objectives, and accomplishments are outlined below.

Criminal Intelligence Service Canada

Criminal Intelligence Service Canada (CISC) is a national body that represents the criminal intelligence efforts of police organizations that maintain a full-time criminal intelligence function. The CISC structure includes a central bureau that is located in Ottawa, as well as nine provincial bureaus. The criminal intelligence requirements of Prince Edward Island are met through the Nova Scotia bureau, while those of the Yukon are served by the British Columbia bureau, and the needs of the Northwest Territories and Nunavut are met by the Alberta and Newfoundland and Labrador bureaus, respectively.

CISC establishes national priorities in its efforts to combat organized crime and launches intelligence projects that are aimed at the various "genres" identified above. Additionally, the CISC has targeted investigations that monitor the sexual exploitation of children, the illegal movement of firearms, and criminal activities associated with Canadian ports.

CISC sponsors an on-line computer database, the Automated Criminal Intelligence Information System (ACIIS), which serves as a national clearing house for intelligence purposes. This database is accessible by all CISC members who, in turn, cooperate in the collection, collation, evaluation, analysis, and dissemination relevant to criminal intelligence matters.

Finally, CISC publishes an annual report designed to provide background on the activities and accomplishments of the organization. It also offers a substantial amount of information and education on the magnitude of the threat of organized crime for the Canadian public and communicates the strategic efforts being taken by Canada's law enforcement agencies against this threat.

Canadian Association of Chiefs of Police, Organized Crime Committee

As part of the committee structure of the Canadian Association of Chiefs of Police (CACP), there is an Organized Crime Committee with the following mandate:

- Governed by the imperative of protecting public safety, security and the quality of life of all citizens of Canada and their communities, the CACP Organized Crime Committee undertakes to lead and strengthen cooperation and coordination amongst law enforcement agencies in the fight against organized crime.

- The committee proposes to invite, evaluate, and promote innovative law enforcement initiatives against organized crime through leadership at both national and international levels, through public communications, awareness and education, and through advocacy with regard to policy and legislation.

- Through strategic decisions guided by information and intelligence from the greater law enforcement community and beyond, the committee will promote policy development and action against organized crime.

- This committee is determined to forge partnerships and model action plans that will offer guidance in a unified law enforcement response to the threat of organized crime in Canada.

The committee includes representatives of the following agencies:

- Calgary Police Service
- Ontario Provincial Police
- Royal Canadian Mounted Police
- Canada Customs and Revenue Agency
- Royal Newfoundland Constabulary
- Organized Crime Agency of British Columbia
- Winnipeg Police Service
- Sûreté du Québec
- Canadian Security Intelligence Service
- Correctional Service Canada
- Halifax Regional Police
- Hamilton Police Service
- Service de Police, Ville de Montréal
- Vancouver Police Department

Nathanson Centre for the Study of Organized Crime and Corruption

The Nathanson Centre for the Study of Organized Crime and Corruption (York University) is based at the Osgoode Hall Law School in Toronto. The centre aims to make a significant contribution to the ongoing study of matters pertaining to the control of organized crime and corruption, and to offer educational programs geared to these phenomena. The centre has a detailed research program that includes the empirical study of organized crime and corruption and encourages an interdisciplinary approach to the issues raised in this area of concern.

The centre is linked with an international network of research institutions and experts in the public, private, and not-for-profit arenas. The centre maintains a detailed website that provides academics, practitioners, and students with a wealth of background information on the nature of organized crime and innovative research and studies dealing with effective enforcement and containment strategies.

Organized Crime Agency of British Columbia

The Organized Crime Agency of British Columbia (OCABC) was created as part of a response to the recommendations of the report of the Organized Crime Independent Review Committee in 1998. The OCABC receives its authority from the *Police Act* (British Columbia) and became fully operational in 2000. It is governed by a board of directors appointed by the Solicitor General of British Columbia and receives most of its funding from the province; however, its work is supported by the federal government through the secondment of Royal Canadian Mounted Police (RCMP) officers. The OCABC has articulated the following major strategic issues with respect to organized crime:

- *Police resource limitations.*   To address the problems of organized crime, police organizations will have to prioritize their own activities and seek additional funding.

- *Globalization of organized crime.*   As the legitimate economy sees a growing trend toward globalization through such enterprises as the World Trade Organization, organized crime is pursuing international opportunities that ignore national boundaries.

- *Fusion of criminal groups into joint ventures.*   The structures of many organized criminal groups mimic those of large private or public corporations. They have their own sophisticated management controls and resources at their disposal to pursue their criminal objectives.

- *Flexibility in multi-commodity criminal activity.*   The diversity of areas susceptible to criminal activity has grown along with the increasing complexity of modern life. Organized criminals are taking advantage of new technologies to pursue the drug trade, credit card fraud, illegal gambling, and computer crimes.

- *Use of advanced technology.*   While advances in information and communications technology have been of enormous benefit to legitimate business and government operations, they have also provided powerful tools for organized crime, including sophisticated methods for concealing the nature and extent of its activities.

- *Lack of public awareness of organized crime.*   Police organizations must attend to the ongoing task of educating and informing the public about their vulnerability to various forms of organized crime. (Adapted from *Organized Crime Agency of British Columbia service plan fiscal years 2002/03 to 2004/05*)

It has been recommended by several policing organizations in Canada that important amendments be made to the *Criminal Code,* the *Corrections and Conditional Release Act,* and the *Income Tax Act of Canada.* These amendments would work to secure longer sentences for those proven to be members of an organized criminal group. It is also necessary to ensure that the assets acquired through such criminal activity are forfeited. Groups like the Canadian Police Association have strongly urged the federal government to reconsider the elimination of the Ports Police and address the weakening of our nation's security at these points of entry.

COMPUTER CRIME

The growing challenge of computer crime has demanded a coordinated and technically competent response from Canadian police organizations. With the exponential proliferation of computers and computer systems around the world, police agencies are called upon to ensure that public systems are secure from misuse and criminal activity. Equally, police organizations must ensure that their own computer systems are protected from invasion, corruption, or damage. We will address this aspect in the following section; however, it is important to remember that police organizations themselves are often the target of computer criminals.

The CACP has recently established an Electronic Crime committee that plans to take a particular interest in the questions of computer crime and work toward the development of strategies for dealing with this particular strain of criminal activity. This committee has the following mandate:

To establish a leadership role in the development of an administrative policy and standards for technology-based investigations, including the promotion of inter-agency cooperation in the detection and investigation of internet-based crime, the establishment of training standards and the identification of effective cooperation strategies to combat E-Crime at local, provincial, national and international levels. (Source: CACP website, **www.cacp.ca**)

One international organization that has been particularly active in the study of computer crime is the Australasian Police Research Centre. This body has produced a wealth of information and background research that is highly relevant to all police organizations struggling with the challenges of e-crime and computer crimes, generally (Morrison, 1990). Because the threat of computer-based crime essentially knows no boundaries, it is imperative that police organizations work on a multi-national, multi-disciplinary basis to confront this growing problem (Thompson, 1990). One important area of concern for police organizations relates to the development of investigative skills that will allow police officers to undertake the complex inquiries required for this particular form of criminal activity. Investigators will need to maintain a high level of sophistication with regard to technological developments as well as be fully aware of the intricacies of relevant legislation pertaining to computer crime in its many manifestations.

INFORMATION AND COMMUNICATIONS TECHNOLOGY

Closely linked to the topic of computer crime, outlined above, is the subject of information and communications technology (ICT) as it relates to police organizations in Canada. There was a time when the most sophisticated piece of equipment in a police station was its radio dispatch system. Today, the situation has changed dramatically, and modern police organizations across Canada are called upon to ensure that they have up-to-date and flexible ICT systems that will allow them to function effectively within the technological environment they are called upon to protect.

At all levels of policing in Canada, there are groups that are dedicated to the challenge of ensuring that their organizations are kept abreast of the latest ICT developments and that they have the wherewithal to function in a highly collaborative, cooperative, and coordinated manner. The CACP has established an "Informatics" committee that seeks to clearly identify priorities and coordinate the association's activities with respect to information management and information technology. This committee also monitors emerging technologies to anticipate possible implications and applications for police operations aimed at the improvement of policing. The members of this committee collaborate very closely with other police stakeholders, including the Canadian Police College, RCMP National Police Services, and the Canadian Police Research Centre.

Within the realm of information technology, there are several developments that police organizations must evaluate and assess as to their potential impact on future criminal activity. It is important that police organizations work to ensure the integrity of their own databases that contain an enormous amount of sensitive information. Furthermore, it is essential that police services provide advice and assistance to other government departments in the investigation and detection of criminal activity involving public databases, computer systems, and networks. Police organizations around the world face competing issues as they consider the new information and communications technology environment. The adoption of numerous legislative and technological methods for ensuring personal privacy has probably occurred because more information about individuals and their lifestyles is available than ever before. This enhanced privacy generally serves the public well, but it makes the search for information about offenders more difficult for police. Rees (2000) says that

> The widespread adoption of machines and devices in the home or office that are capable of connecting to and sharing information with, or even controlled by, other devices throughout the globe may add to the level of business and domestic efficiency. While the concomitant increase in the level of information available about what people are doing may potentially be a valuable source of information for investigating police, such systems are open to abuse and present criminals with opportunities for extortion or blackmail. *(p. iii)*

PUBLIC VERSUS PRIVATE POLICING

A topic of considerable importance over the last decade concerns the respective roles of public and private policing. The number of private police, in both Canada and the United States, has grown substantially over this period of time and shows no signs of abatement. Accordingly, there is considerable debate within the policing community regarding appropriate role clarification for these two forms of policing. On one hand, public policing organizations, including civilian governing authorities, police chiefs, and police associations, argue that they have a clear and compelling mandate to provide the broadest range of policing services within a legislated framework that includes a high degree of public accountability and transparency. On the other hand, private policing representatives argue that they are able to apply sound business practices and economies of scale that offer a viable and cost-efficient alternative to extremely expensive public service departments that are not sufficiently flexible to the safety and security needs of the community.

In the face of this ongoing debate, the Law Commission of Canada (2002) has prepared a detailed discussion paper dealing with the main issues that relate to private versus public policing. Furthermore, the Law Commission of Canada is the sponsor of a major international conference that facilitates dialogue and debate on the issues that are central to this matter. This conference, entitled "In Search of Security: an international conference on policing and security," will generate some valuable insights in this important area and will advance innovative thinking about new public and community safety models that will see a blending of these two modes of policing.

All of the major policing stakeholders have prepared positions with respect to these issues, and there is a wealth of research to enlighten those interested in this topic. For example, the Canadian Police Association has developed an interesting presentation on the topic of private security in Canada. Furthermore, during its last annual conference the association passed the following resolution with respect to private security:

> Be it resolved that the Canadian Police Association and its member organizations are opposed to the downloading of public law responsibilities and granting of police authority to private investigators and security guards. (Source: CPA website, **www.cpa-acp.ca**)

PROTECTION OF PRIVACY AND FREEDOM OF INFORMATION

Police and public safety executives in Canada must ensure that they work to balance two frequently competing priorities as they pursue their organizational goals and objectives. One priority relates to their responsibility to ensure the protection of individual privacy. This means that the information about an individual must be secured and used only for appropriate institutional purposes authorized by law. The other priority relates to their equally compelling responsibility to adhere to the principle of freedom of information. This relates closely to the concept of accountability and transparency in the conduct of public affairs. Police organizations, as a category of public service, must not operate under a cloak of secrecy without there being a compelling justification for such an approach. Increasingly, police organizations in Canada are called upon to operate with a much greater degree of openness and candour than ever before. These legitimate, but competing, priorities constitute one of the major challenges facing policing executives today.

Across Canada, there are agencies, boards, and commissions that are entrusted with responsibility for ensuring that citizens are not denied access to information that they have a right to receive and review. Also, there are formal bodies established that will protect individuals and groups from unwarranted and unwelcome invasions of privacy by public or private corporations, including police organizations. At the federal level, the Privacy Commissioner of Canada functions to protect the privacy rights of individuals and is active in fighting government programs that seek to erode such privacy for the sake of security. Recently, the Privacy Commissioner of Canada, supported by many other provincial and territorial information and privacy commissioners, has publicly criticized the federal government's Bill C-17 (*Public Safety Act*), which gives the RCMP broad powers to use passenger information for general law enforcement purposes. The Privacy Commissioner has also stated a concern about the plans of the Canada Customs and Revenue Agency (CCRA) to assemble a substantial surveillance database that would monitor Canadians and others travelling to Canada. This opposition clearly highlights the delicate balance that exists between those concerned about public safety and security and those dedicated to personal freedom and liberty.

It is worthwhile recalling that all levels of government in Canada have some form of freedom of information and protection of privacy legislation in place. The existence of such legislation has placed a considerable onus on police organizations to ensure that their own policies, procedures, and practices are fully consistent with the requirements of these statutes. What follows is a highly selective sampling of recent information and privacy cases that have involved various police departments in Canada. The following examples should serve to highlight the critical importance of these bodies, as well as the interdependence of policing organizations and administrative tribunals in Canada.

Privacy Commissioner of Canada and RCMP Video Surveillance

In this case, the federal Privacy Commissioner responded to a complaint filed by the Information and Privacy Commissioner of British Columbia with respect to the proposed installation of surveillance cameras by the RCMP in the downtown core of the City of Kelowna. Following the Privacy Commissioner's detailed review of this issue, which concluded in October 2001, it was determined that continuous video surveillance in a public place was an invasion of privacy that could not be justified for public safety and security purposes. The Privacy Commissioner found that the RCMP's planned video surveillance was in contravention of the *Privacy Act.*

Alberta Information and Privacy Commissioner and the Edmonton Police Service

In March 2002, the Alberta Information and Privacy Commissioner issued an order (Order F2002-001) dealing with a complaint against the Edmonton Police Service (EPS). The EPS had issued a news release to the media outlining a criminal investigation with respect to the complainant. Included in this release was a photograph of the complainant and details about his activities. The complaint alleged that the EPS released private information to the media and, thereby, caused trauma and injury to the complainant's entire family. Following a thorough review of the issues and legislation pertinent to this case, the commissioner determined that the release of personal information about this person by the EPS was in compliance with the *Freedom of Information and Protection of Privacy Act.*

Alberta Information and Privacy Commissioner and the Calgary Police Commission

In this case, reported on July 6, 2000, the commissioner received a complaint from the president of the Calgary Police Association alleging that a member of the Calgary Police Commission had breached the privacy rights of the members of the association executive by sending business correspondence to their home addresses. This matter was investigated and reviewed by the commissioner, who offered a number of recommendations to ensure that the privacy of the association members was respected and that procedures be put in place to avoid future problems of this nature.

Information and Privacy Commissioner of Ontario and the Ministry of Public Safety and Security

This case relates to a reconsideration order (PO-2063-R) by the commissioner with respect to an earlier interim order (PO-2033-I) issued by that office in August 2002. It deals with a request under the *Freedom of Information and Protection of Privacy Act* from a member of the media for access to "all video footage recorded by the Ontario Provincial Police (OPP) at Ipperwash Provincial Park (Ipperwash) from September 5–7, 1995… [and] all photos taken by the OPP at Ipperwash Provincial Park from September 5–7, 1995." This case was highly controversial, and the commissioner undertook a complex examination of the issues and legalities pertaining to the request. Accordingly, the commissioner ordered the Ministry of Public Safety and Security to disclose several records (i.e., photographs) on the basis that they did not contain "personal information." Furthermore, the commissioner upheld the ministry's decision to deny access to other categories of records (other photographs) on the basis that they did contain "personal information."

Information and Privacy Commissioner of Ontario and the Ottawa Police Services Board

This case pertains to a newspaper reporter's request to the Ottawa Police Services Board for a copy of the investigation report prepared by the department's Professional Standards Section dealing with an incident involving an Ottawa city councillor who was also a member of the Ottawa Police Services Board. The board denied access to the records at issue, and the commissioner conducted a thorough review of the matter. As a result of this review, the commissioner found that the board had indeed released a public statement on the incident involving the councillor in question, which included a detailed summary of the occurrence, as well as an explanation of the discussions and conclusions of the board. The commissioner ruled that the public interest had been served by these releases and accordingly upheld the decision of the board.

POLICE EDUCATIONAL REFORM

If police organizations across Canada are going to be able to maintain the position of leadership that they have acquired, it is essential that a great deal of attention be paid to the issue of police educational reform. It is essential that police organizations develop a strategic approach to the identification, development, and continuous support of future police leaders.

Police Leadership Program (Ontario Association of Chiefs of Police)

The Ontario Association of Chiefs of Police has partnered with the Joseph L. Rotman School of Management at the University of Toronto to create an innovative Police Leadership Program that will help ensure a future generation of qualified police executives. Recognizing a long-standing need for high-quality police leadership, it introduces participants to the broad range of organizational principles and practices that will help them meet the specialized challenges associated with policing in Canada. The program has the following objectives:

- develop participants' leadership competencies, particularly in visioning, strategic thinking, and organizational development;

- sharpen their understanding of the political, social, and economic forces that shape the external environment within which the police services operate;

- develop their skills in dealing with politicians, media, and members of community groups and interest groups; and

- develop their personal leadership and management abilities.

The courses within this program are designed to reinforce fundamental concepts and to instruct program participants in competent executive-level skills. Program instruction is based on the effective team-based learning philosophy that is the hallmark of executive programs delivered by the Rotman School of Management. Courses combine in-class instruction, peer-based learning, simulations, role-playing, team projects, and individual assignments. The program also includes guest speakers representing the sectors of law enforcement, government, and the general public.

The Canadian Police College, Police Executive Centre

The Canadian Police College in Ottawa works to address the challenges facing police organizations from a learning perspective. Presently, the college sponsors a Police Executive Centre that seeks to offer specialized learning programs for this audience. The centre brings together small groups of participants and attempts to generate discussions that will meet their personal, professional, and organizational learning needs. The following topics provide a focus for these workshop sessions:

- strategic business planning

- organized crime

- knowledge management

- communicating like a leader

- critical thinking skills

- globalization and human rights

- police labour relations

- policing and politics

- internal and external communications

Ontario Police College

Established in 1962, the Ontario Police College (OPC) is a major centre for police training, education, and development in the province. It offers a comprehensive range of courses, seminars, workshops, and learning programs designed to contribute to the continuous improvement of police personnel in Ontario's diverse communities.

An element within the organizational structure of the Policing Services Division of the Ministry of Public Safety and Security, OPC follows a detailed business plan that is consistent with the overall strategic directions of that ministry. OPC develops programs that are designed to ensure that high standards of professionalism, innovation, and accountability are observed. During 1999, the college provided learning programs for some 7,760 students and continues to demonstrate a strong commitment to excellence in police training.

Saskatchewan Police College, University of Regina

Affiliated with the University of Regina, the Saskatchewan Police College collaborates with municipal police services in the province to provide ongoing training, education, and development programs that support professional policing services. The college applies a systems approach to the research, design, delivery, and evaluation of its learning programs.

Justice Institute of British Columbia

The Justice Institute of British Columbia was established in 1978 to offer learning programs across the province. Among the groups that attend the institute for training, education, and development are the following:

- police officers, counsellors, correctional workers, and court staff responsible for offenders and victims;

- firefighters, paramedics, search and rescue personnel, and other individuals (both paid and volunteer) engaged in emergency response;

- persons involved with mediation, negotiation, and conflict resolution; and

- community agency staff and other social service providers.

Recently, the institute graduated students who completed their new Justice and Public Safety Leadership degree program. This program is a partnership with Simon Fraser University and is designed for adults at mid-career who wish to acquire an undergraduate degree on a part-time basis while pursuing full-time employment. Culminating in a Bachelor of General Studies, the program recognizes work experience and other post-secondary learning. It includes the following required courses:

- collective bargaining

- organizational theory

- organizational behaviour

- negotiation and conflict resolution

- research methods

- crime and deviance

- advanced university writing

- writing theory and practice

- ethical decision making

- critical thinking and the humanities

- leadership

- gender and social issues

Another innovation in student learning involves a partnership between the Justice Institute of British Columbia and Royal Roads University to deliver a Master of Arts degree in justice and public safety leadership. It aims to provide graduates with the requisite leadership and management skills and abilities to function as effective professionals within the realm of justice and public safety. The program begins with an intensive period of residency at the Justice Institute of British Columbia and proceeds to course work completed through an Internet-based distance learning approach. Another period of residency involves an action research project that will allow students to focus on a specific work-related issue leading to meaningful organizational change.

Canadian Association of Police Educators

The Canadian Association of Police Educators (CAPE) is a national group of individuals committed to the promotion of excellence in the realm of police learning. CAPE works to encourage dialogue on problems and challenges in the area of police education and to develop "best practices" in this area of activity. The association undertakes research leading to innovative approaches to police learning and hosts an annual conference for that purpose.

Ontario Association of Police Educators

The Ontario Association of Police Educators (OAPE) seeks to develop a higher degree of knowledge, skills, and abilities within the discipline of policing. The OAPE promotes standardized training methods and works to facilitate an ongoing exchange among those who are involved with police training, education, and development across the province. The association undertakes research in certain relevant areas and works to promote innovative approaches and learning initiatives.

The OAPE also sponsors a conference that attempts to bring together police education practitioners in support of the association's mandate.

POLICE OVERSIGHT AND CIVILIAN GOVERNANCE

Another significant element in the arena of public policing involves the whole question of police oversight and civilian governance. This area has been studied with considerable interest (Hann et al., 1985; Stenning, 1981; and McKenna, 2003) and continues to provoke controversy and concern. As policing in Canada continues to evolve, it is essential that the debate over appropriate mechanisms and methodologies for police oversight and civilian governance be sustained. There are many signs that progress is being made on this front as police leaders, public officials, and social critics come together to better understand and articulate their various positions in an open and constructive manner.

There continues to be considerable controversy over the operation of a high-profile civilian oversight entity, like the Special Investigations Unit in Ontario. However, many signs suggest that a wide range of policing and public safety stakeholders are becoming comfortable with a regime of accountability and responsibility that ensures the police remain a public service that is closely supervised and scrutinized in the public interest.

The work of the Canadian Association of Civilian Oversight of Law Enforcement (CACOLE) is relevant to anyone interested in civilian oversight in Canada. Also, the national efforts of the Canadian Association of Police Boards (CAPB) provide a valuable window on the world of civilian governance of policing. These representative bodies work to ensure that the delivery of policing services throughout the country is stable and sound. The role of civilian overseers and police governing authorities is, in essence, to provide an enlightened stewardship for the public interest in policing and public safety. They represent those who guard the guardians and they carry an enormous public trust.

CONCLUSION

This chapter has turned our attention to the most pressing issues pertaining to public safety, security, and surveillance. It is certainly true that these critical issues will deeply influence the evolution and development of public policing in Canada. Therefore, it is essential that students and practitioners of policing have a clear understanding of such matters and come to a high level of awareness regarding their impact.

The complex subject of terrorism has been touched upon in a most preliminary manner. It is clear that the response of Canada's public police organizations to the threat of terrorism will have a profound effect on the shape of our domestic models of policing. There is an apparent tension between a security model of policing that supports a more militaristic approach and a proactive, or problem-solving, approach that is directed at the local dimension of public safety and supports community wellness and quality of life. By becoming aware of the fundamental distinctions between these diametrically opposed approaches to the threat of terrorism, we will be better able to formulate the appropriate response that will accord with our democratic principles and obligations.

This chapter has also dealt in some detail with the long-standing problem of organized crime. This is another broad topic with many facets that continue to challenge the ingenuity and intelligence of Canada's police organizations. It is also an area that requires considerable coordination and cooperation between various police jurisdictions and across municipal, provincial, national, and international boundaries. This is an area of police activity that can only be successful when there is ongoing strategic planning and prolonged tactical engagement of organized crime networks.

Some of the issues that relate to information and communications technology have been addressed in this chapter. This area represents an unprecedented challenge to police competence in so far as it introduces an element of technical sophistication that most police organizations have never had to possess in the past. As information and communications technology develops, it brings with it increasing opportunities for illegal activity. It also introduces significant risks for all manner of organizations, both public and private, with respect to the protection of their data and proprietary information. Canadian police organizations must not only be able to ensure the integrity of their own files, information, and systems, but must be prepared to protect, investigate, and substantiate charges against those who would threaten the integrity of information and communications systems. Furthermore, this chapter speaks to the dual responsibility of police organizations to adhere to the goals of freedom of information as well as the responsibility for the protection of privacy.

In the context of ongoing, continuous learning, this chapter has included some consideration of the most significant innovations in Canadian police educational reform. This is an area that should see increased academic and professional attention as police organizations come to the full realization that their most important resources are their people. Police personnel, both uniformed and civilian, should be seen as knowledge workers within their organizations and should be provided with the training, education, and development necessary to pursue that role.

Finally, this chapter considers some of the key aspects of police oversight and civilian governance in Canada. This represents another area of study that will continue to evolve as police organizations become increasingly more accountable to the public they serve and move toward appropriate levels of transparency and ethical practice.

REFERENCES

Australasian Police Commissioners. Electronic Crime Steering Committee (2001). *Electronic crime strategy of the Police Commissioners' Conference, Electronic Crime Steering Committee 2001–2003*. Payneham, South Australia : Australasian Centre for Policing Research.

Beare, Margaret (2000). *Facts from fiction: tactics and strategies of addressing organized crime and organized criminals*. Presentation at the Canadian Police College seminar series: "Perspectives on organized crime in Canada."

Beare, Margaret and R.T. Naylor (1999). *Major issues relating to organized crime: within the context of economic relationships: prepared for the Law Commission of Canada*. Toronto : Nathanson Centre for the Study of Organized Crime and Corruption, York University.

Canadian Intelligence Service Canada (2002). *Annual report on organized crime in Canada*. Ottawa : Canadian Intelligence Service Canada.

Canadian Police Association (2000). *Brief to the House of Commons sub-committee on organized crime*. Ottawa : Canadian Police Association.

Cavoukian, Ann (1999). *Biometrics and policing: comments from a privacy perspective*. Toronto : Information and Privacy Commissioner Ontario.

Etter, Barbara (2002). *Critical issues in hi-tech crime*. Payneham, South Australia : Australasian Centre for Policing Research.

Hann, Robert G. et al. (1985). "Municipal police governance and accountability in Canada: an empirical study." *Canadian Police College Journal,* Vol. 9, No. 1, pp. 1–85.

Kaku, M. (1998). *Visions: how science will revolutionise the 21st century and beyond*. Oxford : Oxford University Press.

Law Commission of Canada (2002). *In search of security: the roles of public police and private agencies*. Ottawa : The Law Commission of Canada.

McKenna, Paul F. (1999). *Foundations of policing in Canada*. Toronto : Prentice-Hall Canada.

——— (2000). *Foundations of community policing in Canada*. Toronto : Prentice-Hall Canada.

——— (2003). *Police powers II*. Toronto : Pearson Education Canada.

Morrison, P. (1990). *Computer crime: the improvement of investigative skills — final report part one*. Payneham, South Australia : Australasian Centre for Policing Research (Report series; no. 96).

Organized Crime Agency of British Columbia (2002). *Organized Crime Agency of British Columbia service plan fiscal years 2002/03 to 2004/05*. Vancouver, B.C. : Organized Crime Agency of British Columbia

PricewaterhouseCoopers (2001). *Strategic human resources analysis of public policing in Canada*. Ottawa : Canadian Association of Chiefs of Police and Canadian Police Association.

Rees, Andrew (2000). *Technology environment scan: a technology forum report*. Payneham, South Australia : Australasian Centre for Policing Research (Report series; no. 133.1).

Smith, R. (1998). *Criminal exploitation of new technologies*. Canberra : Australian Institute of Criminology (Trends and issues in crime and criminal justice; no. 93).

Stenning, Philip C. (1981). *Police commissions and boards in Canada*. Toronto : Centre of Criminology, University of Toronto.

Sterling, C. (1995). *Crime without frontiers*. New York : McGraw Hill.

Thompson, D. (1990). *Computer crime: the improvement of investigative skills: final report part two: issues for the development of a law enforcement strategy*. Payneham, South Australia : Australasian Centre for Policing Research (Report series; no. 101).

WEBLINKS

Australasian Centre for Policing Research

www.acpr.gov.au

Canadian Association of Police Educators

www.cpc.gc.ca/cape_e.htm

Canadian Intelligence Service Canada

www.cisc.gc.ca

Canadian Police College, Police Executive Centre

www.cpc.gc.ca/execut_e.htm

Justice Institute of British Columbia

www.jibc.bc.ca

Nathanson Centre for the Study of Organized Crime and Corruption

www.yorku.ca/nathanson

Ontario Association of Chiefs of Police, Police Leadership Program, Joseph Rotman School of Management (University of Toronto)

www.oacp.on.ca/programs/leadership.html

Ontario Association of Police Educators

www.oape.org

Ontario Police College

www.mpss.jus.gov.on.ca/english/police_serv/ont_police_col/overview.html

Organized Crime Agency of British Columbia

www.ocabc.org

Royal Roads University

www.royalroads.ca

Society for the Policing of Cyberspace

www.polcyb.org

QUESTIONS FOR CONSIDERATION
AND DISCUSSION

1. How have the events of September 11, 2001, altered the approach to public safety and security in Canada?

   _____

   _____

   _____

   _____

   _____

   _____

   _____

   _____

2. Which of the "genres" of organized crime identified in this chapter poses the most serious threat to public safety and security in Canada? Explain your answer.

   _____

   _____

   _____

   _____

   _____

   _____

   _____

3. How have recent innovations in information and communications technology (ICT) complicated the operation and administration of police services in Canada?

   _____

   _____

   _____

_____

_____

_____

_____

_____

_____

4. What new technologies do you believe will be most significant for the advancement of policing in Canada? Explain your answer.

_____

_____

_____

_____

_____

_____

_____

_____

_____

5. What, if any, are the possible dangers that may be associated with the continued technological sophistication of modern police services in Canada?

_____

_____

_____

_____

_____

_____

_____

_____

_____

6. Does the private security industry require standards of conduct and operation that are similar to those applied to police?

_____

_____

_____

_____

_____

_____

7. Are the current regulatory regimes sufficient to control the private security industry?

_____

_____

_____

_____

_____

_____

8. Is communication and cooperation between public police and private security agencies appropriate?

_____

_____

_____

_____

_____

_____

9. What limits should be placed on the ability of police organizations to secure information on possible criminal suspects as part of their investigative mandate?

_____

_____

_____

_____

_____

_____

10. What categories and classes of information should police organizations be able to protect from public access to ensure the safety of their officers and the confidentiality of sensitive investigative techniques?

_____

_____

_____

_____

_____

_____

11. What is the ideal model for the design, development, delivery, and evaluation of police education and learning programs?

_____

_____

_____

_____

_____

_____

_____

_____

_____

_____

_____

RELATED ACTIVITIES

1. Examine current newspapers and journals in search of articles and items that deal with the police response to terrorism. Compare the response to international terrorism with the domestic security concerns of local police services.

2. Review the following report prepared by Statistics Canada: _Organized crime activity in Canada: a pilot survey of 16 police services_ (May 1999). This report includes quantitative information about criminal organizations that is drawn from police services across Canada: Calgary, Edmonton, Halifax, Toronto, Montreal, Ontario, RCMP, Sudbury, Sûreté du Québec, Vancouver, Waterloo, Winnipeg, Hamilton, Regina, and the Service de police de Québec. The study includes five categories of

criminal organizations in Canada: outlaw motorcycle gangs, Asian-based organized crime gangs, Italian-based organized crime gangs, Aboriginal-based crime gangs, and Eastern European-based crime groups.

3. Review the *Organized Crime Agency of British Columbia's Service Plan fiscal years 2002/03 to 2004/05* (see links to this document on the agency's website) and consider how this organization may be effective in combating the presence of organized crime within that jurisdiction. Consider how an agency of this nature might benefit other provincial jurisdictions in Canada.

4. Review the discussion paper issued by the Law Commission of Canada on public police and private agencies and consider the issues that are raised by that document.

5. Make a few visits to police departments in your area and determine what their policies, procedures, and protocols are with regard to organized crime.

6. Review the programs offered by the various organizations and institutions in Canada that are committed to police education and compare their approaches to this challenge.

chapter ten

The Future of the Canadian Political Realm and Public Administration

Recently the more clear-sighted of our ruling classes have recognized that progress is a more complex matter than was envisaged by those who had believed that a better society would arise ineluctably from technology . . . . We are now faced with easily calculable crises (concerning population resources, pollution, etc.) which have been consequent upon the very drive to mastery itself. The political response to these interlocking emergencies has been a call for an even greater mobilisation of our technological representation of reality. More technology is needed to meet the emergencies which technology has produced.

—George Grant, *Technology and Justice*

Chief Bill Closs, Kingston Police.

Kingston Police

Learning Objectives

1. Identify aspects relating to professional ethics that are likely to affect the future of Canadian politics and public administration.

2. Outline the major issues that will affect citizen participation and engagement in the future of Canada.

3. Identify aspects of public accountability that will influence the future of politics and public administration in Canada.

4. List the key features of current developments that are likely to influence the future of the Canadian federation.

5. Outline the fundamental issues that concern the Aboriginal peoples of Canada and demonstrate how they are linked to the future of Canadian politics and public administration.

INTRODUCTION

This chapter, in attempting to cover a wide range of complex, strategic topics, will only begin to trace the outline of these areas of inquiry. It will be important for the reader to pursue various avenues of research to become more fully conversant with the meaning and magnitude of these topics.

This final chapter will discuss the growth in concern over professional ethics as it relates to the future of Canadian politics and public administration. The discussion will include an introductory review of the ethical guidelines that have been proposed for our country's parliamentarians. It will also make a connection with the evolving world of police ethics.

The significant topic of citizen engagement will be reviewed in this section, and some consideration will be given to various developments in the area of citizen participation in the political process. Here we will touch upon the troubling matter of voter apathy and the impact this is likely to have on the future of politics and public administration in Canada.

Under the broad heading of accountability, we will examine the serious question of the "democratic deficit" as it has been labeled in recent political discussions. We will look at how police organizations, as an example of a highly accountable public service, often fail to fully acquit themselves of their deepest responsibilities to the public by their very adherence to rules and regulations. Also, we will explore how administrative secrecy may have a detrimental effect on the future of our political realm, posing a threat to transparent public administration.

The related topics of restorative justice and dispute resolution will be touched on as important future trends that are likely to have an impact on government and public administration.

Next, we will look at the various facets of the continuing federation within Canada. There are several threats on the political horizon, and it is important to reference them in any presentation on the health of our body politic.

Finally, we will look at certain issues pertaining to Aboriginal peoples and their quest for an independent justice system. A great deal is happening in this area, including the negotiation of Indian land claims and the work of several organizations representing the specific interests of various tribal councils. In every province and territory changes are taking place, or are under consideration, that speak to the sovereignty of First Nations, Métis, or Inuit peoples.

PROFESSIONAL ETHICS

Following a storm of controversy over the conduct of certain members of the Chrétien government, new ethical guidelines for parliamentarians were introduced by the Liberal Party. The release of a proposed *Code of Conduct for Parliamentarians* came closely upon the heels of the departure of the federal Solicitor General, Lawrence MacAulay, whose alleged patronage activities had been reviewed and criticized by the federal Ethics Commissioner. This new code is intended to

- maintain and promote public confidence and trust in the integrity of parliamentarians as well as the respect and confidence that society places in Parliament as an institution;

- demonstrate to the public that parliamentarians are held to standards that place the public interest ahead of their private interests and to provide a transparent system by which the public may judge this to be the case;

- provide for greater certainty and guidance for parliamentarians in how to reconcile their private interests with their public duties and functions; and

- foster consensus among parliamentarians by establishing common standards and by providing the means by which questions relating to proper conduct may be answered by an independent, non-partisan adviser. (Source: *Code of Conduct for Parliamentarians,* 2002)

In addition to the proposed *Code of Conduct for Parliamentarians,* the Liberal government has introduced a draft bill for the establishment of an ethics commissioner, reporting directly to Parliament. This official would be responsible for the administration of the code of conduct and would provide advice to the Prime Minister on the *Code of Interest and Post-Employment Code for Public Office Holders.*

The Canadian Association of Chiefs of Police (CACP) has established an Ethics Subcommittee as part of the CACP's Human Resources Committee. Over the last few years, the CACP has invested considerable time and effort in the formulation of an "Ethical Framework" that has been outlined in detail elsewhere (McKenna, 2003). To assist its membership in taking this ethical framework seriously, the CACP Ethics Subcommittee has articulated a strategic plan for its implementation. The CACP hopes that members will make the ethical framework an integral component of their day-to-day practice as police leaders. Of course, the value of any such approach to ethical behaviour is only as good as its actual application. Therefore, to ensure that the ethical framework serves as a genuine guide to police executives across the country, it is essential that a formal evaluation process be rigorously applied to determine if it has truly been successful in meeting its goals and objectives.

Police Accountability and Victimization by Procedure

Over the past 15 years, police organizations have enshrined the philosophy of community policing in their officers and communities. Fundamental strategies included community partnerships to identify and resolve victimization issues and other quality of life concerns. We have also been keen to market our efforts to the public. One can understand the immediate buy-in from the community when the police commit to openly sharing information, and promise to actively listen and to respond to the community through mutual trust building. The police were ensured success as we promoted the new philosophy.

What many police agencies overlooked was that we were actually setting in motion a movement (the community) that would come to expect, and demand, a higher benchmark in police accountability.

By way of example, a police organization will answer a call that requires a high-risk police takedown, which is the epitome of police power in action. It involves professionally trained police officers pointing their firearms, emergency lights flashing, and police officers shouting orders. It is not uncommon to find that at the conclusion of this tactical maneuver the suspects were innocent. Most often, there is no wrongdoing on the part of the police because they acted on reliable information and carried out the takedown procedure professionally and carefully. As for the innocent persons, they were in the wrong place at the wrong time, when suddenly the police crashed into their lives and, for a few intense minutes, treated them like criminal suspects in accordance with police training. Another more common example is apparent when we recognize that police are in the business of apprehending criminals, and not a day goes by when they don't interview, interrogate, suspect, or accuse an innocent person of wrongdoing.

It is no longer acceptable for the police to simply explain the situation by proclaiming, "We were responding to a call and acted in good faith." The new accountability that comes with these scenarios is that police agencies must broaden their understanding of victimization. Traditionally, police have believed that a victim was a person who has been injured in some emotional, physical, or material way. We recognize the victim of a motor vehicle collision, of an assault, or of a theft. The new era of police accountability has brought recognition and understanding that law enforcement activities can create victims. Police officials must realize and accept that, by virtue of their extraordinary statutory powers, citizens can be victimized even when police actions are legal.

During a 37-year policing career, 11 years of which were directly involved in police training, I have never read or heard a police organization admit that police procedures, even when legal and justified, can victimize persons. Such an admission is long overdue.

Police investigative techniques instruct officers not to narrow the search for the suspect too quickly, to prevent preconceived ideas of guilt that result in police continuing, as a normal course of doing business, to accuse innocent persons. Therefore we must adjust our organizational mindset and training to ensure adequate response to the needs of victims, including those victimized by justifiable police procedures by officers acting in good faith. It's moving community policing and police accountability to a higher level. It's also police accepting responsibility for a sacred trust given to us by the community.

—Chief Bill Closs, Kingston Police

CITIZEN ENGAGEMENT

The whole question of citizen engagement in the various mechanisms of democracy has come under scrutiny over the last several decades. Many observers have begun to question the strength of our adherence to our fundamental democratic institutions. In Chapter 4 ("The Political Process"), we looked at voter participation in the electoral process. Frequently, the voter turnout statistics are disappointing, if not alarming. It is not uncommon for political analysts to speculate about the causes of apparent voter apathy, and it is frequently suggested that people have begun to lose faith in our institutions of government. For example, in 1997 an Angus Reid poll discovered that 47 percent of the Canadian public indicated a lack of confidence in our courts, while 54 percent expressed a lack of confidence in the prison system, and 74 percent noted a lack of confidence in the *Young Offenders Act*. To address these concerns, it is essential for decision-makers at various levels of government to focus on the challenge of citizen engagement.

Beyond simple participation in the process of voting, citizens should be activated to truly connect with the processes, programs, and policies of government. There is a saying that you get the government you deserve. It would appear that many Canadians are coming to the realization that we should be working to deserve better government. This will certainly come about if individuals, groups, and organizations find ways to reconnect with the responsibilities and obligations that necessarily coincide with the rights and liberties we enjoy under our existing form of government.

ACCOUNTABILITY AND ADMINISTRATIVE SECRECY

We expect our government officials, both elected and bureaucratic, to be accountable to the citizens they serve. We also expect our government institutions to function in ways that are open to scrutiny and that operate in the public interest. However, many political commentators and potential leadership candidates such as Paul Martin, Jr., are highlighting the problem of the "democratic deficit." This term has taken on a great deal of importance in Canadian politics and will be debated heatedly in the coming years.

What this term implies is that our democratic institutions have begun to erode and we are seeing a tendency for power to become overly concentrated within the central agencies of government, such as the federal Cabinet. This of course means that locally elected Members of Parliament (MPs) find themselves in a difficult position where they are not able to fully represent the interests of their constituencies. In fact, these MPs are frequently expected to adhere without question to the policy directions of the party executive. While concern over a perceived "democratic deficit" is not unique to Canada (Avineri, 2001; Moravcsik, 2002), when the power of the executive becomes overwhelmingly strong, questions arise about the independence of our parliamentary institutions. As citizens, we should be alarmed when a central government undertakes policy initiatives that have far-reaching socio-economic consequences for the Canadian public without substantial and meaningful dialogue or debate. Much of the growing social activism in Canada is driven by citizens' groups concerned about the activities of the World Trade Organization, the International Monetary Fund, and other globalizing forces that appear to have removed fundamental political power from the citizenry. Organizations like the Council of Canadians, Amnesty International, the Canadian Civil Liberties Association, and many others have worked to call attention to the dangers of such forces.

A recent appointment to the Supreme Court of Canada has been criticized as being too secretive (Ziegel, 2002). While it is true that the Prime Minister is not obligated to consult with Parliament, the provinces, or the public on this kind of appointment, there is opposition to the secrecy and lack of transparency in this matter. Concerns about the lack of accountability and the growing tendency to invoke a privilege of administrative secrecy have created a deep sense of distrust among Canadians. The future of our political systems depends substantially on how these concerns are addressed by our political leaders.

RESTORATIVE JUSTICE AND DISPUTE RESOLUTION

The related topics of restorative justice and dispute resolution have become increasingly important in many quarters of government in Canada. For example, the government of Nova Scotia has embraced the concept and principles of restorative justice and has made considerable headway in applying those principles to its policy framework. At the prompting of the governments of Canada and Italy, the United Nations Economic and Social Council (the Council) endorsed the basic principles of restorative justice in July 2002. Furthermore, the Council adopted a resolution encouraging all countries to apply basic principles on the use of restorative justice programs in criminal matters. In the realm of policing, the Royal Canadian Mounted Police (RCMP) have incorporated a restorative justice model in their community policing programs.

There is also a growing interest in dispute resolution practices and techniques in Canada, and many law schools across the country are beginning to offer courses in this approach to conflict resolution. In many areas of criminal justice, scholars are working out alternative approaches to the justice system and reconciliation (Bianchi, 1994).

Restorative Justice

Restorative justice involves a systematic approach to crime that places an emphasis on the need to provide healing for the victim and the community, as well as the perpetrator of the crime. It recognizes that healing approaches must be applied to truly resolve the conditions that cause crime in our society. It also attempts to replace an adversarial approach to justice with one that is restorative for the victim, the community, and the offender. Programs that apply the principles of restorative justice typically include the following elements:

- identification of steps to repair the harm done;
- involvement of all stakeholders; and
- transformation of the traditional relationship between communities and their governments.

The Nova Scotia Restorative Justice Program

The Nova Scotia Department of Justice's restorative justice program provides an interesting counterpoint to the traditional criminal justice system. Beginning in 1997, the province formed a multi-disciplinary committee to work on the development of a suitable model for this program. The provincial government is to establish a legal framework and standards and ensure that the program is monitored.

In Nova Scotia, the restorative justice program provides intake opportunities at the following entry points across the criminal justice system:

Police entry point. This occurs at the pre-charge stage and involves referral by police officers.

Crown entry point. This occurs at the post-charge stage and involves referral by Crown attorneys.

Court entry point. This occurs at the post-conviction or pre-sentence stage and involves referral by judges.

Corrections entry point. This occurs at the post-sentence stage and involves referrals by correctional services or victims' services staff.

This approach is experimental and requires effective evaluation to determine the degree to which it has been successful in addressing the needs of communities for justice and fairness.

Dispute Resolution

Dispute resolution involves the use of any appropriate process for the resolution of disputes. Many techniques can be applied in this area, including administrative, judicial, or legislative decision-making approaches such as arbitration, mutual problem-solving, consensus-building, negotiation, facilitation, mediation, and conciliation. These processes typically involve a "third party" acting as an objective and impartial participant in the process.

Federal Dispute Resolution Services

In 1992, the federal Department of Justice established a program area known as Dispute Resolution Services (DRS). The creation of this program area clearly demonstrates the government's commitment to dispute resolution as a valid approach to dealing with conflict in ways that are less adversarial. Accordingly, the DRS provides negotiation and facilitation services and has team members who are trained in offering neutral assessments, arbitration, mediation, and other dispute resolution techniques. While dispute resolution is not seen as a complete replacement for the litigation process normally found in our courts of law, it is offered as a highly flexible alternative that may be effectively used in resolving interpersonal conflicts in ways that are fully satisfactory to the parties involved. It is not difficult to imagine the many applications alternative approaches to dispute resolution might have in areas of policing and law enforcement.

ADR Institute of Canada

The ADR Institute of Canada (ADR Canada) was established as a non-profit, national centre for leadership in the promotion and advancement of dispute resolution approaches. With seven affiliates across the country, it provides support and representation to a multi-disciplinary range of professionals offering dispute resolution services. ADR Canada offers the following services:

- a clearing house of information on dispute resolution;
- access to a range of tools and techniques for practitioners and users of dispute resolution;
- national standards and a code of ethics for trainers in ADR; and
- national accreditation for mediators and arbitrators.

THE CONTINUING FEDERATION

The future of Canada is an ongoing concern for academics, politicians, policy-makers, and the public-at-large. We have seen serious threats to the integrity of the Canadian federation during the FLQ crisis of the 1970s, the emergence of a powerful sovereignty movement in Quebec politics, and the growth of official political parties that run on platforms that express a high degree of regional alienation. The Bloc Québécois, the erstwhile Reform Party (under Preston Manning), and the current Canadian Alliance are all signs that the status quo within Confederation is not fully acceptable. Within a context of continuous change it is impossible to predict what the configuration of our federal government will be in the next 10, or 20, years. However, it is certain that there will be dramatic, disturbing, and deeply disruptive matters to deal with as we arrive at whatever accommodations are necessary to ensure political, social, and economic stability. What follows are a few of the current undertakings that will likely contribute to the configuration of change within Canada:

- Royal Commission on Renewing and Strengthening Our Place in Canada
- Commission on the Future of Health Care in Canada
- Kyoto Protocol to the United Nations Framework Convention on Climate Change

Royal Commission on Renewing and Strengthening Our Place in Canada

Beginning its work in June 2002, the Newfoundland and Labrador Royal Commission on Renewing and Strengthening Our Place in Canada turned its attention to examining the more than five decades of the province's participation in Canadian Confederation. In addition to a series of public consultations, the commissioners have sponsored a program of research that will review the following topical areas pertinent to the province's involvement in the Canadian family:

- expectations at Confederation and progress achieved;
- how Newfoundland and Labrador is viewed in Canada;
- terms of union and other constitutional provisions;
- contributions of Newfoundland and Labrador to Canada;
- arrangements with Canada that hamper prosperity and self-reliance;
- demographic changes and impact on prospects for youth;
- how Newfoundland and Labrador can take advantage of its strategic location; and
- challenges of the global economy.

Commission on the Future of Health Care in Canada

On November 28, 2002, the Commission on the Future of Health Care in Canada, under former Saskatchewan premier Roy Romanow, tabled its final report to the Canadian people. This report, while not directly affecting policing and public safety concerns, will certainly have a profound influence over federal-provincial relations as they pertain to the funding and features of our future health care system. This in turn will affect the rapidly aging Canadian population, and any significant demographic initiative is certain to create policy issues for our policing services. Furthermore, the central importance of health care to the quality of life of Canadians and the enormous human and financial resources required to sustain any national health care infrastructure will have a decidedly deep impact on the nature of our Canadian federation.

Kyoto Protocol to the United Nations Framework Convention on Climate Change (Kyoto Accord)

The Kyoto Accord is based on a United Nations convention that was adopted in 1992 to address the problems associated with global climate change. While removed from the normal concerns of policing and public safety, this accord is having a significant impact on the relationship between the federal government and its provincial counterparts. The province of Alberta was particularly vocal in its opposition to the original terms of the Kyoto Accord, and it is likely that this agreement will continue to be the focus of considerable debate and controversy within the Canadian federation. In December 2002 the federal Liberal government ratified the accord and began proceeding with its implementation. Many provincial governments have serious reservations about the impact the agreement will have on their economy, their resources, and their industries. They have been, therefore, reluctant to offer immediate support to the federal government on this front. What this debate will do to the stability of Canadian confederation is unknown; however, it is certain to create unrest and conflict within the context of intergovernmental relations.

ABORIGINAL PEOPLES AND AN INDEPENDENT JUSTICE SYSTEM

There are several initiatives relating to our Aboriginal peoples (including First Nations, Métis, and Inuit peoples) that have the potential to significantly affect the future of the Canadian polity. One of the most important relates to the ongoing deliberations pertaining to Indian land claims. The work of the Indian Claims Commission is both pivotal and fundamental to the nature of our relationship as a country with our Aboriginal peoples. The resolution of the residential schools crisis will speak volumes to the quality of the relationship we are able to forge with those who have been damaged by the presence of the non-Aboriginal peoples. The movement toward a nation-to-nation relationship between the various Aboriginal peoples and the Canadian government is powerfully felt across the country and, of course, includes within its ambit, the struggle for an independent justice system for Aboriginal peoples, both on and off reserves. The Royal Commission on Aboriginal Peoples presented its final report to the House of Commons in November 1996. That report (in five volumes), offered a number of long-range recommendations pertaining to the relationship between Aboriginal and non-Aboriginal peoples in Canada. In the realm of justice and policing, the following topics continue to be of major importance:

- community and restorative justice;
- legislation and legal issues;
- policing and corrections; and
- residential schools.

The federal government has sponsored an Aboriginal Canada website portal which provides an access point to a wealth of information and research on matters pertinent to the ongoing relationship between Aboriginal and non-Aboriginal peoples across Canada.

The Policing of First Nations Communities

In May 2000, the First Nations Chiefs of Police Association released a report on the first phase of a projected seven-part examination of human resources within First Nations policing in Canada. This report, *Setting the context: the history of First Nations policing in Canada,* provides a review of the development of police services in Canada's First Nations communities, including the growth of non-Aboriginal policing. This study, undertaken in partnership with Human Resources Development Canada sets the stage for necessary change and alteration in the model of First Nations policing across the country.

The Aboriginal Justice Learning Network

The Aboriginal Justice Learning Network (AJLN) includes representatives from the Canadian justice system and various Aboriginal communities. The AJLN is coordinated through an office in the Department of Justice and works toward the administration and provision of a broad range of justice services for Aboriginal peoples. Established in 1996, the AJLN is guided by an Advisory Committee that is essentially composed of Aboriginal peoples. The committee is supported by individuals with professional backgrounds in law, community work, policing, the judiciary, and youth issues. The Aboriginal Justice Strategy pursued by the committee has three objectives:

- to support Aboriginal communities as they take greater responsibility for the administration of justice;

- to help reduce crime and incarceration rates in the communities that administer justice programs; and

- to improve Canada's justice system to make it more responsive to the justice needs and aspirations of Aboriginal people. (Source: Department of Justice Canada, Aboriginal Justice Learning Network website, **http://canada.justice.gc.ca/en/ps/ajln/strat.html** [accessed November 30, 2002])

CONCLUSION

This chapter makes an effort to briefly sketch an outline of some of the critical issues that will affect the future reality of Canada's political landscape and its institutions of public administration. This presentation is in no way exhaustive, and the reader should supplement it by making reference to the various resources and websites listed at the end of this chapter.

In considering the ethical dimension of Canadian politics and public service, there is clearly much to be done to ensure that the theoretical frameworks that have been painstakingly developed become part of the working mentality of our parliamentarians and civil servants. The future health of the Canadian polity will hinge quite significantly on the capacity of these two categories of public officials to internalize the ethical codes of conduct that have been formulated for their roles as stewards of the public good.

The notion of citizen engagement is also critical to the future of Canadian politics and public administration. Without a significant degree of citizen participation in the processes, practices, and programs of government, it is unlikely that we will continue to enjoy our current rights and freedoms. Citizen responsibility, rightly understood, is the price that must be paid for such freedoms. It is relevant that the success, or failure, of the approach to community, or problem-oriented, policing in Canada hinges mightily upon the capacity of communities to become involved and engaged in a genuine partnership with their police services. Democratic principles and institutions cannot be sustained in a climate of neglect, benign or otherwise.

Accountability is a hallmark of our liberal democratic traditions and structures. To sustain appropriate levels of accountability, it is important to ensure that those who conduct business in the public interest are aware of their roles and responsibilities. These people must know that there are trustworthy police guardians who are qualified and dedicated to the oversight function, who will offer authentic service on behalf of the public. This reality applies to all branches and levels of government, and across all jurisdictions in Canada.

The emergence of restorative justice and the growing interest in various approaches to dispute resolution are positive signs of evolution in our approach to justice and fairness. While these trends are still relatively young in their application in Canada, they hold considerable promise for the future and will continue to have an important impact on our political institutions and in our approach to public administration.

Canada's future is necessarily uncertain. We cannot know precisely what will happen to this unique federation as we move more deeply into the 21st century. However, this chapter has outlined several pivotal issues that will doubtless affect our evolution as a country.

Finally, the long-standing question of the future of our Aboriginal peoples will profoundly affect the future of our Canadian polity. Much is happening on this vast horizon and the points of contact with our political institutions and our public affairs are almost unlimited. The issues of justice, fairness, sovereignty, and human rights are all inextricably linked with this area of concern. Canada's future cannot be read without rewriting the chapter that addresses our Aboriginal peoples.

REFERENCES

Avineri, Shlomo (2001). "A democratic deficit." *Jerusalem Post,* October 22, 2001.

Bianchi, Herman (1994). *Justice as sanctuary: toward a new system of crime control.* Bloomington : Indiana University Press.

Bowman, James S. (1994). *Ethical frontiers in public management: seeking new strategies for resolving ethical dilemmas.* San Francisco : Jossey-Bass.

Cooper, Terry L. (1998). *The responsible administrator: an approach to ethics for the administrative role. 4th ed.* San Francisco : Jossey-Bass.

First Nations Chiefs of Police Association (2000). *Setting the context: the policing of first nations communities: module one.* Ottawa : First Nations Chiefs of Police Association in partnership with Human Resources Development Canada.

Kernaghan, Kenneth (1991). "Managing ethics: complementary approaches." *Canadian Public Administration,* Vol. 34, No. 1 (Spring), pp. 132–145.

Law Commission of Canada (2000). *Restoring dignity: responding to child abuse in Canadian institutions.* Ottawa : Law Commission of Canada.

Lewis, Carol W. (1991). *The ethics challenge in public service: a problem-solving guide.* San Francisco : Jossey-Bass.

McKenna, Paul F. (2003). *Police powers II.* Toronto : Pearson Education Canada.

Moravcsik, Andrew (2002). *In defense of the "democratic deficit": reassessing legitimacy in the European Union.* Cambridge, Mass. : Center for European Studies, Harvard University (Center for European Studies working paper; no. 92).

Nova Scotia. Department of Justice (1998). *Restorative justice: a program for Nova Scotia.* Halifax : Nova Scotia Department of Justice.

Royal Commission on Renewing and Strengthening Our Place in Canada (2002). *Royal Commission on Renewing and Strengthening Our Place in Canada consultation document: September 4, 2002.* St. John's : The Royal Commission.

Segal, Hugh (2001). "Policing and democratic values: how is Canada doing?: notes for an address to the Canadian Police College seminar series." Ottawa : Institute for Research on Public Policy.

———— (2002). "Police and politicians: the accountability/independence conundrum: notes for an address to the Canadian Police College executive seminar." Ottawa : Institute for Research on Public Policy.

Senge, Peter M. (1990). "The leader's new work: building learning organizations." *Sloan Management Review,* Vol. 12, No. 1 (Fall), pp. 7–23.

Ziegel, Jacob (2002). "A Supreme democratic deficit." *National Post,* Monday, August 12, 2002.

WEBLINKS

Aboriginal Canada Portal

www.aboriginalcanada.gc.ca

Aboriginal Justice Learning Network

http://canada.justice.gc.ca/en/ps/ajln/index.html

Access to Justice Network

www.acjnet.org

ADR Institute of Canada (ADR Canada)

www.adrcanada.ca

Canada. Royal Commission on Aboriginal Peoples

www.ainc-inac.gc.ca/ch/rcap/rpt

Canadian Association of Chiefs of Police, Human Resources Committee, Ethics Subcommittee

www.cacp.ca

Commission on the Future of Health Care in Canada

www.healthcarecommission.ca

Dispute Resolution Services, Department of Justice Canada

http://canada.justice.ca/en/ps/drs

First Nations Chiefs of Police Association

www.soonet.ca/fncpa

Guide to Federal Initiatives for Urban Aboriginal People

www.pco-bcp.gc.ca/docs/Publications/abor/guide_e.html

Indian Claims Commission

www.indianclaims.ca

**Kyoto Protocol to the United Nations
Framework Convention on Climate Change**

http://unfccc.int/resource/convkp.html

Nova Scotia Department of Justice, Restorative Justice Program

www.gov.ns.ca/just/rj/rj-contents.htm

**Restorative Justice Online (International
Centre for Justice and Reconciliation)**

www.restorativejustice.org

**Royal Commission on Renewing and
Strengthening Our Place in Canada**

www.gov.nf.ca/royalcomm

QUESTIONS FOR CONSIDERATION
AND DISCUSSION

1. How will new ethical rules for parliamentarians impact public confidence in our elected officials?

_____

_____

_____

_____

_____

_____

_____

2. Is the ethical framework adopted by the Canadian Association of Chiefs of Police relevant at all levels of policing in Canada?

_____

_____

_____

_____

_____

_____

3. How will greater citizen engagement in the political and policy-making processes improve policing and public safety in Canada?

_____

_____

_____

_____

_____

_____

_____

4. What degree of administrative secrecy is warranted within a liberal democratic country like Canada?

_____

_____

_____

_____

_____

_____

_____

5. To what extent should the various levels of government be held accountable to the public? Is it appropriate for police organizations to have a higher standard of accountability than other public services? Explain your answer.

_____

_____

_____

_____

_____

_____

_____

6. Are the principles and practices of restorative justice and dispute resolution applicable to the complete range of activities and services offered by the various levels of government?

_____

_____

_____

_____

_____

_____

_____

7. What are the most serious threats to the stability of the Canadian federation? Explain your answer.

   _____

   _____

   _____

   _____

   _____

   _____

   _____

   _____

8. Which threats to the stability of the Canadian federation are most likely to have an impact on policing and public safety?

   _____

   _____

   _____

   _____

   _____

   _____

   _____

   _____

9. How would an independent Aboriginal justice system interact with the existing policing and public safety services in the rest of Canada?

   _____

   _____

   _____

   _____

   _____

   _____

   _____

   _____

10. What difficulties or problems can you foresee with respect to an independent Aboriginal justice system in Canada?

_____

_____

_____

_____

_____

_____

_____

_____

RELATED ACTIVITIES

1. Examine the proposed *Code of Conduct for Parliamentarians* and compare that document to the Ethical Framework developed by the Canadian Association of Chiefs of Police. Consider the similarities and differences between these two ethical models.

2. Review the various major commissions and inquiries that have been undertaken recently and consider their potential impact on the future of policing in Canada. Examples could include:

   • Royal Commission on Renewing and Sustaining Our Place in Canada

   • Royal Commission on Aboriginal Peoples

   • Commission on the Future of Health Care in Canada

3. Research the topic of restorative justice and consider its application to the Canadian criminal justice system.

4. Examine selected issues relating to justice and policing that address the concerns of Canada's Aboriginal peoples.

CANADA ACT 1982 including the CONSTITUTION ACT, 1982
1982, c. 11 (U.K.)

[29th March 1982]

[Note: The English version of the Canada Act 1982 is contained in the body of the Act; its French version is found in Schedule A. Schedule B contains the English and French versions of the Constitution Act, 1982.*]*

An Act to give effect to a request by the Senate and House of Commons of Canada.

Whereas Canada has requested and consented to the enactment of an Act of the Parliament of the United Kingdom to give effect to the provisions hereinafter set forth and the Senate and the House of Commons of Canada in Parliament assembled have submitted an address to Her Majesty requesting that Her Majesty may graciously be pleased to cause a Bill to be laid before the Parliament of the United Kingdom for that purpose.

Be it therefore enacted by the Queen's Most Excellent Majesty, by and with the advice and consent of the Lords Spiritual and Temporal, and Commons, in this present Parliament assembled, and by the authority of the same, as follows:

1. *The Constitution Act, 1982* set out in Schedule B to this Act is hereby enacted for and shall have the force of law in Canada and shall come into force as provided in that Act.

2. No Act of the Parliament of the United Kingdom passed after the *Constitution Act, 1982* comes into force shall extend to Canada as part of its law.

3. So far as it is not contained in Schedule B, the French version of this Act is set out in Schedule A to this Act and has the same authority in Canada as the English version thereof.

4. This Act may be cited as the Canada Act, 1982.

SCHEDULE B
Constitution Act, 1982
PART I

CANADIAN CHARTER OF RIGHTS AND FREEDOMS   Whereas Canada is founded upon principles that recognize the supremacy of God and the rule of law:

Guarantee of Rights and Freedoms

1. The Canadian Charter of Rights and Freedoms guarantees the rights and freedoms set out in it subject only to such reasonable limits prescribed by law as can be demonstrably justified in a free and democratic society.

Fundamental Freedoms

2. Everyone has the following fundamental freedoms:
   (a) freedom of conscience and religion;
   (b) freedom of thought, belief, opinion and expression, including freedom of the press and other media of communication;
   (c) freedom of peaceful assembly; and
   (d) freedom of association.

Democratic Rights

3. Every citizen of Canada has the right to vote in an election of members of the House of Commons or of a legislative assembly and to be qualified for membership therein.

4. (1) No House of Commons and no legislative assembly shall continue for longer than five years from the date fixed for the return of the writs at a general election of its members.
   (2) In time of real or apprehended war, invasion or insurrection, a House of Commons may be continued by Parliament and a legislative assembly may be continued by the legislature beyond five years if such continuation is not opposed by the votes of more than one-third of the members of the House of Commons or the legislative assembly, as the case may be.

5. There shall be a sitting of Parliament and of each legislature at least once every twelve months.

Mobility Rights

6. (1) Every citizen of Canada has the right to enter, remain in and leave Canada.
   (2) Every citizen of Canada and every person who has the status of a permanent resident of Canada has the right
      (a) to move to and take up residence in any province; and
      (b) to pursue the gaining of a livelihood in any province.

(3) The rights specified in subsection (2) are subject to

    (a) any laws or practices of general application in force in a province other than those that discriminate among persons primarily on the basis of province of present or previous residence; and

    (b) any laws providing for reasonable residency requirements as a qualification for the receipt of publicly provided social services.

(4) Subsections (2) and (3) do not preclude any law, program or activity that has as its object the amelioration in a province of conditions of individuals in that province who are socially or economically disadvantaged if the rate of employment in that province is below the rate of employment in Canada.

Legal Rights

7. Everyone has the right to life, liberty and security of the person and the right not to be deprived thereof except in accordance with the principles of fundamental justice.

8. Everyone has the right to be secure against unreasonable search or seizure.

9. Everyone has the right not to be arbitrarily detained or imprisoned.

10. Everyone has the right on arrest or detention

    (a) to be informed promptly of the reasons therefor;

    (b) to retain and instruct counsel without delay and to be informed of that right; and

    (c) to have the validity of the detention determined by way of *habeas corpus* and to be released if the detention is not lawful.

11. Any person charged with an offence has the right

    (a) to be informed without unreasonable delay of the specific offence;

    (b) to be tried within a reasonable time;

    (c) not to be compelled to be a witness in proceedings against that person in respect of the offence;

    (d) to be presumed innocent until proven guilty according to law in a fair and public hearing by an independent and impartial tribunal;

    (e) not to be denied reasonable bail without just cause;

    (f) except in the case of an offence under military law tried before a military tribunal, to the benefit of trial by jury where the maximum punishment for the offence is imprisonment for five years or a more severe punishment;

    (g) not to be found guilty on account of any act or omission unless, at the time of the act or omission, it constituted an offence under Canadian or international law or was criminal according to the general principles of law recognized by the community of nations;

    (h) if finally acquitted of the offence, not to be tried for it again and, if finally found guilty and punished for the offence, not to be tried or punished for it again; and

    (i) if found guilty of the offence and if the punishment for the offence has been varied between the time of commission and the time of sentencing, to the benefit of the lesser punishment.

12. Everyone has the right not to be subjected to any cruel and unusual treatment or punishment.

13. A witness who testifies in any proceedings has the right not to have any incriminating evidence so given used to incriminate that witness in any other proceedings, except in a prosecution for perjury or for the giving of contradictory evidence.

14. A party or witness in any proceedings who does not understand or speak the language in which the proceedings are conducted or who is deaf has the right to the assistance of an interpreter.

Equality Rights

15. (1) Every individual is equal before and under the law and has the right to the equal protection and equal benefit of the law without discrimination and, in particular, without discrimination based on race, national or ethnic origin, colour, religion, sex, age or mental or physical disability.

   (2) Subsection (1) does not preclude any law, program or activity that has as its object the amelioration of conditions of disadvantaged individuals or groups including those that are disadvantaged because of race, national or ethnic origin, colour, religion, sex, age or mental or physical disability.

[Note: This section became effective on April 17, 1985. See subsection 32(2) and the note thereto.]

Official Languages of Canada

16. (1) English and French are the official languages of Canada and have equality of status and equal rights and privileges as to their use in all institutions of the Parliament and government of Canada.

   (2) English and French are the official languages of New Brunswick and have equality of status and equal rights and privileges as to their use in all institutions of the legislature and government of New Brunswick.

   (3) Nothing in this Charter limits the authority of Parliament or a legislature to advance the equality of status or use of English and French.

17. (1) Everyone has the right to use English or French in any debates and other proceedings of Parliament.

   (2) Everyone has the right to use English or French in any debates and other proceedings of the legislature of New Brunswick.

18. (1) The statutes, records and journals of Parliament shall be printed and published in English and French and both language versions are equally authoritative.

   (2) The statutes, records and journals of the legislature of New Brunswick shall be printed and published in English and French and both language versions are equally authoritative.

19. (1) Either English or French may be used by any person in, or in any pleading in or process issuing from, any court established by Parliament.

   (2) Either English or French may be used by any person in, or in any pleading in or process issuing from, any court of New Brunswick.

20. (1) Any member of the public in Canada has the right to communicate with, and to receive available services from, any head or central office of an institution of

the Parliament or government of Canada in English or French, and has the same right with respect to any other office of any such institution where

(a) there is a significant demand for communications with and services from that office in such language; or

(b) due to the nature of the office, it is reasonable that communications with and services from that office be available in both English and French.

(2) Any member of the public in New Brunswick has the right to communicate with, and to receive available services from, any office of an institution of the legislature or government of New Brunswick in English or French.

21. Nothing in sections 16 to 20 abrogates or derogates from any right, privilege or obligation with respect to the English and French languages, or either of them, that exists or is continued by virtue of any other provision of the Constitution of Canada.

22. Nothing in sections 16 to 20 abrogates or derogates from any legal or customary right or privilege acquired or enjoyed either before or after the coming into force of this Charter with respect to any language that is not English or French.

Minority Language Educational Rights

23. (1) Citizens of Canada

(a) whose first language learned and still understood is that of the English or French linguistic minority population of the province in which they reside, or

(b) who have received their primary school instruction in Canada in English or French and reside in a province where the language in which they received that instruction is the language of the English or French linguistic minority population of the province, have the right to have their children receive primary and secondary school instruction in that language in that province.

(2) Citizens of Canada of whom any child has received or is receiving primary or secondary school instruction in English or French in Canada, have the right to have all their children receive primary and secondary school instruction in the same language.

(3) The right of citizens of Canada under subsections (1) and (2) to have their children receive primary and secondary school instruction in the language of the English or French linguistic minority population of a province

(a) applies wherever in the province the number of children of citizens who have such a right is sufficient to warrant the provision to them out of public funds of minority language instruction; and

(b) includes, where the number of those children so warrants, the right to have them receive that instruction in minority language educational facilities provided out of public funds.

Enforcement

24. (1) Anyone whose rights or freedoms, as guaranteed by this Charter, have been infringed or denied may apply to a court of competent jurisdiction to obtain such remedy as the court considers appropriate and just in the circumstances.

(2) Where, in proceedings under subsection (1), a court concludes that evidence was obtained in a manner that infringed or denied any rights or freedoms guaranteed by this Charter, the evidence shall be excluded if it is established that, having regard to all the circumstances, the admission of it in the proceedings would bring the administration of justice into disrepute.

General

25. The guarantee in this Charter of certain rights and freedoms shall not be construed so as to abrogate or derogate from any aboriginal, treaty or other rights or freedoms that pertain to the aboriginal peoples of Canada including

    (a) any rights or freedoms that have been recognized by the Royal Proclamation of October 7, 1763; and

    (b) *any rights or freedoms that may be acquired by the aboriginal peoples of Canada by way of land claims settlement.*

    (c) any rights or freedoms that now exist by way of land claims agreements or may be so acquired.

[Note: Paragraph 25(b) (in italics) was repealed and the new paragraph substituted by the Constitution Amendment Proclamation, 1983 *(No. 46 infra).]*

26. The guarantee in this Charter of certain rights and freedoms shall not be construed as denying the existence of any other rights or freedoms that exist in Canada.

27. This Charter shall be interpreted in a manner consistent with the preservation and enhancement of the multicultural heritage of Canadians.

28. Notwithstanding anything in this Charter, the rights and freedoms referred to in it are guaranteed equally to male and female persons.

29. Nothing in this Charter abrogates or derogates from any rights or privileges guaranteed by or under the Constitution of Canada in respect of denominational, separate or dissentient schools.

30. A reference in this Charter to a province or to the legislative assembly or legislature of a province shall be deemed to include a reference to the Yukon Territory and the Northwest Territories, or to the appropriate legislative authority thereof, as the case may be.

31. Nothing in this Charter extends the legislative powers of any body or authority.

Application of Charter

32. (1) This Charter applies

    (a) to the Parliament and government of Canada in respect of all matters within the authority of Parliament including all matters relating to the Yukon Territory and Northwest Territories; and

    (b) to the legislature and government of each province in respect of all matters within the authority of the legislature of each province.

    (2) Notwithstanding subsection (1), section 15 shall not have effect until three years after this section comes into force.

[Note: This section came into force on April 17, 1982. See the proclamation of that date (No. 45 infra).]

33. (1) Parliament or the legislature of a province may expressly declare in an Act of Parliament or of the legislature, as the case may be, that the Act or a provision thereof shall operate notwithstanding a provision included in section 2 or sections 7 to 15 of this Charter.

    (2) An Act or a provision of an Act in respect of which a declaration made under this section is in effect shall have such operation as it would have but for the provision of this Charter referred to in the declaration.

    (3) A declaration made under subsection (1) shall cease to have effect five years after it comes into force or on such earlier date as may be specified in the declaration.

    (4) Parliament or the legislature of a province may re-enact a declaration made under subsection (1).

    (5) Subsection (3) applies in respect of a re-enactment made under subsection (4).

Citation

This Part may be cited as the *Canadian Charter of Rights and Freedoms*.

Index